Climate Change and Carbon Markets

CLIMATE CHANGE AND CARBON MARKETS

Elisa Gomez Gonzalez

Delve
PUBLISHING

www.delvepublishing.com

Climate Change and Carbon Markets

Elisa Gomez Gonzalez

Delve Publishing
224 Shoreacres Road
Burlington, ON L7L 2H2
Canada
www.delvepublishing.com

Email: orders@arclereducation.com

ISBN: 978-1-77469-523-4 (Hardcover)

ABOUT THE AUTHOR

Born in the Canary Islands, Elisa Gomez has always been aware of the immense potential and limitations that resources provide to communities. Following her passion towards the environment and sustainable development mechanism, she completed the Madrid Environmental Sciences degree and later completed a Master's degree in International Sustainable Development and Corporate Sustainability. She has worked for a wide range of industries and sectors as a sustainability consultant and analyst implementing international standards, achieving certifications, and ensuring the management processes were effective. The sectors englobe oil and gas, construction, telecommunications, and software design. Moved by a great passion towards her work, she decided to quit the corporate environment and set up her own company, Better Together (www.bettertogether-sustainability.com) an online sustainability company that specializes in simple, effective, and understandable sustainability practices for all types of companies. They focus especially on Startups and SMEs, as these are growing; ensuring that they are sustainable right from the start without the burden of complicated, time-consuming, and expensive measures.

TABLE OF CONTENTS

LIST OF FIGURES

mole fractions for CO_2 in parts per million per year are shown as gray columns (left), CH_4 in parts per billion per year (center), and N_2O in parts per billion per year (right) (Source: WMO Global Atmosphere Watch)

Figure 6.4. 1960–2019 ensemble meantime series and ensemble standard deviation (2-sigma, shaded) of global ocean heat content (OHC) anomalies relative to the 2005–2017 climatology for the 0–300m (gray), 0–700m (blue), 0–2000m (yellow) and 700–2000m depth layer (green). The ensemble mean is an outcome of a concerted international effort, and all products used are listed at the end of the report and in the legend of Figure 6.5. Note that values are given for the ocean surface area between 60°S-60°N, and limited to the 300m bathymetry of each product. Updated from von Schuckmann et al. (2020)

Figure 6.5. Linear trends of global ocean heat content (OHC) as derived from different temperature products (colors). References are provided at the end of the report. The ensemble mean and standard deviation (2-sigma) is given in black, respectively. The shaded areas show trends from different depth layer integrations, i.e., 0–300m (light turquoise), 0–700m (light blue), 0–2000m (purple) and 700–2000m (light purple). For each integration depth layer, trends are evaluated over four periods: historical (1960–2019), altimeter era (1993–2019), golden Argo era (2005–2019), and the most recent period 2010–2019 (Updated from von Schuckmann et al., 2020).

Figure 6.6. (left) Satellite altimetry-based global mean sea level for January 1993 to October 2020 (last data: 13 October 2020). Data from the ESA climate change Initiative sea-level project, from January 1993 to December 2015 (thick black curve); extended by the Copernicus Marine and Environment Service, CMEMS, until August 2020 (blue curve) and with near real-time altimetry data from the Jason-3 mission beyond August 2020 (red curve). The thin black curve is a quadratic function that best fits the data. Vertical dashed lines mark the start of each year from 2016 to 2021. (right) Interannual variability of the (with the quadratic function shown in the left-hand panel subtracted) global mean sea level (blue curve) with the multivariate ENSO index (red curve) superimposed

Figure 6.7. (A) Global map showing the highest MHW category (for definitions, see Marine heatwave data) experienced at each pixel over the course of the year, estimated using the NOAA OISST v2.1 dataset (reference period 1982–2011). Light gray indicates that no MHWs occurred in a pixel over the entire year; (B) Stacked bar plot showing the percentage of ocean pixels experiencing an MHW on any given day of the year; (C) Stacked bar plot showing the cumulative percentage of the ocean that experienced an MHW over the year [24]. Horizontal lines in this figure show the final percentages for each category of MHW; (D) Stacked bar plot showing the cumulative number of MHW days averaged over all pixels in the ocean [25] (Source: Robert Schlegel, IMEV)

Figure 6.8. (left) Surface pH values based on ocean acidification data submitted to the 14.3.1 data portal (http://oa.iode.org) for the time period from 1 January 2010 to 8 January 2020. Grey circles – calculated pH of data submissions (including all data sets with data for at least two carbonate parameters); blue circles – average annual pH

target year. The entire series was revised to reflect new information made available since the publication of the last edition of the report; it replaces all series published previously

Figure 6.16. Displays FAO-WFP early warning analysis of acute food insecurity hotspots in July 2020 [107]. The hotspot countries were selected through a consensus-based analytical process by FAO and WFP, focusing primarily on: levels of food insecurity and malnutrition prior to the COVID-19 pandemic, projections of acute food insecurity, presence of existing compounding shocks and stressors, including climatic conditions, plant pests and animal diseases, economic crisis or political instability, conflict, and insecurity

Figure 6.17. Desert locust movement prediction using FAO´s Desert Locust Information System (DLIS). The unusual weather conditions in fall 2019 along the Arabian Peninsula generated strong cyclones and heavy rains resulting in a higher than normal vegetation growth providing ideal conditions for desert locusts. During the 2020 compounded crisis, early warning and early action response played a key role in saving up to 1.52 million tons of cereals (enough to feed 10 million people in one year), worth around US$456 million, across nine eastern African countries and Yemen [111]. Other anticipatory actions promoted by the FAO include: i) providing key inputs to the most food-insecure farmers, including seeds and agricultural tools to allow timely planting; ii) implementing cash-based programs; iii) distributing animal feed and mineral supplements to vulnerable livestock raisers; iv) monitoring and controlling all infested/breeding areas, enhancing surveillance and conducting control operations to prevent swarms from forming and spreading [112]

Figure 6.18. Impact of climate change mitigation policies. Sources: IMF (2020b); International Energy Agency; OECD Environmental Policy Stringency Index; and Worldwide Patent Statistical Database. Note: In panels c, d, e whiskers represent 90 percent confidence intervals. In panel f shaded areas represent high and low climate sensitivity. CCS = carbon capture and storage. See IMF (2020b) for full details

Figure 6.19. Climate change adaptation policies toolkit

Figure 10.1. Reporting and communication areas

Figure 10.2. Reporting process

Figure 10.3. Materiality analysis example

Figure 11.1. UNGC certifications figures

Figure 11.2. UNGC sustainable development goals

Figure 11.3. ISO 26000 framework

Figure 11.4. GRI framework

LIST OF ABBREVIATIONS

AAUs	assigned amount units
ACE	accumulated cyclone energy
AI	artificial intelligence
AMR	antimicrobial resistance
AO	arctic oscillation
AVISO	archiving, validation, and interpretation of satellite oceanographic data
BOTA	Botswana Training Authority
CCS	carbon catch and capacity
CDM	clean development mechanism
CDSB	Climate Disclosure Standards Board
CER	Certified Emissions Reduction Credit
CLG	Corporate Leadership Group
CMEMS	Copernicus marine environment monitoring service
COATS	CO_2 allowance tracking system
CRM	customer relationship management
EC	European Commission
EEI	earth energy imbalance
EfS	education for sustainability
EIB	European investment bank
ENSO	El Niño southern oscillation
EPIC	embankment project for inclusive capitalism
ERUs	earned reduction units
EU ETS	EU emissions trading system
EU	European Union
EUAs	European Union allowances
EV	electric vehicles

GAW	global atmospheric watch
GCOS	global climate observing system
GCW	global cryosphere watch
GFAS	global fire assimilation system
GOOS	global ocean observing system
GRI	global reporting initiative
GTN	global terrestrial networks
GTN-G	GTN-glaciers
GTN-H	GTN-hydrology
GTN-P	GTN-permafrost
IDPs	individual development plans
INCR	investor network on climate risk
INDCs	intended nationally determined contributions
KPIs	key performance indicators
MGGRA	midwestern greenhouse gas reduction accord
MHWs	marine heatwaves
NAPs	national allocation plans
NASA	National Aeronautics and Space Administration
NDCs	nationally determined contributions
NGOs	Nongovernmental Organizations
NMHS	National Meteorological and Hydrological Monitoring Services
OCG	Observations Coordination Group
OECD	Organization for Economic Co-operation and Development
OISST	optimum interpolation sea surface temperature
OOPC	Ocean Observations Physics and Climate Panel
OTC	over-the-counter
PG&E	Pacific Gas and Electric
RECs	renewable energy credits
RFP	request for proposals
SDGs	sustainable development goals
SMB	surface mass balance
SMEs	small and medium-sized enterprises
SSA	Sub-Saharan Africa
TVET	Technical and Vocational Education and Training

UHC	United Health Care
UNFCCC	UN Framework Convention on Climate Change
WBCSD	World Business Council for Sustainable Development
WCI	Western Climate Initiative
WCRP	World Climate Research Program
WGMS	World Glacier Monitoring Service
WSSD	World Summit on Sustainable Development

SYNOPSIS

This book will run you through the entire process of setting up a sustainability management system, that everybody feels accountable and responsible for, as well as highlighting how it can build resilient mechanisms towards natural disasters.

The first chapters will set up the scene describing how sustainability is currently being implemented, what are the trends and challenges. These initial chapters are very important as they contemplate the main assumptions and characteristics of current practices, which will enable the reader to better understand the measures presented afterward.

The last chapters focus mainly on how sustainability practices can be implemented effectively into different departments within businesses to ensure a good deployment of efforts and allocation of resources.

INTRODUCTION

Sustainability emerged from the need to protect and ensure our environment, our society, and our economic system assumes responsibility and works to cause as little harm as possible. Sustainability as a concept stepped onto the world stage in 1992 at the World's first Earth Summit in Rio. Since then, awareness, and the application of sustainable practices have been on the rise.

This book will include the background information needed to understand sustainability management systems and their integration into businesses fully. The reader will be provided with the benefits, risks, and disadvantages of the different methods to assure comprehension and analytical thinking are stimulated.

Renewable technology development goes hand-in-hand with sustainability management systems. Thus, this book covers the trends, future expectations, and forecasts as well as a thorough overview of what is happening around the world.

A simple methodology will be presented to effectively approach the integration of sustainability in all types of businesses. Case studies, practical examples, and templates will be provided in each phase of the methodology, enabling the reader to assess and implement sustainability as they read along. Relevant supporting information will be provided at the end of each chapter on specific topics to take into consideration.

Promoting and empowering people into incorporating sustainability is my personal mission. I strongly believe in the positive impact it provides and in the added value that it enables for all. Sustainability does not have to be complicated, expensive, or unrelated to the business activity. The numerous benefits its incorporation provides to all types of companies are always worth the journey.

In the end, the main goal of sustainability, in my humble opinion, is to do and use business as a force for good.

EXECUTIVE SUMMARY

Sustainable business examples are progressively showing up on the radar of organizations around the globe because of current trends and communication efforts. An effective company or association is very much reliant on its investor's needs and expectations. Investors are interested in sustainability, impact, and added value.

A reasonable association meets the desires of diverse partners. A genuine feasible association is capable of bringing together monetary, social, and natural contemplations. Such businesses are effectively ready to meet the desires of both internal and external partners, for example, investors, clients, government, employees, and groups to guarantee long-term success for all.

In any case, most businesses neglect to execute feasible objectives. Regularly, businesses neglect to execute sustainable business practices because of insufficient reconciliation with business methodology, absence of help from authority, constrained comprehension, and inclusion of workers at all levels, and additionally, nonappearance of frameworks and procedures that enable a systematize support.

Sustainability can enable businesses to prevail with regard to current economic business trends. Albeit numerous businesses dedicate themselves to sustainability and frame a strategy, its implementation, and coordination depend upon effectively embracing the following:

- Empowering and driving economic business practices at all levels of the association.
- Creating effective frameworks and procedures.

PREFACE

Sustainability management systems have proven not only to benefit the company that applies them but also to the world as a whole. Sustainability is slowly but steadily being incorporated into public policies, governments, private sector companies, and even freelancers. The world has begun to believe that we can actually operate and perform in a way that is respectful to the environment, society, and the economy enabling us to develop sustainably.

This is not merely a trend or the hot topic for the coming years, but it is a crucial aspect of we want to continue prospering a thriving. A great part of sustainable management practices relies on people. People that work with a purpose; feel engaged, accountable, and responsible for their actions. People that want to make a difference for good. These people together form teams that work with purpose, feel engaged, connected, accountable, and responsible for their actions. This group of people forms companies, businesses, organizations, and overall a community of people that have in mind to be sustainable and to provide the best for future generations.

The purpose is the main driver for millennials in choosing a workplace and a career. This shift from other generational trends implies that companies and the workplace industry globally need to face a transformation. Millennials which will shortly constitute the biggest generation in the workplace have a broader understanding of sustainability. Not only a broader understanding but they go one step further and actually reject working for companies that do not create value and have a positive impact where they operate. Their understanding of sustainable development together with their immense technological capacity, nowadays, set the perfect scene for all types of companies and organizations to start implementing sustainability management systems.

With sustainability management systems, renewable technology development goes hand-in-hand. Thus, along with an overview of what is happening around the world, this book covers the trends, future expectations, and forecasts.

Typically, sustainability management systems or sometimes referred to as corporate social responsibility (CSR) are very much dependent upon international standards, regulation, legislation, and certifications which are very much industry-related or country-related. This means that they are heavily influenced by the culture, the economic positioning in the market, and the country's legal and political trends.

Moreover, globalization has enabled access to products and services where they were not available before. All of these changes and limitations have created a difficult scenario for companies, where implementing sustainability measures implies allocating a great part of the budget and result in very time-consuming for employees.

The main aim of this book is to provide simple, effective, understandable, and applicable mechanisms for all readers. To enable and empower them to implement management systems that are adapted to the reality and expectations of their business.

In this book, you will find all of those mechanisms and resources needed to ensure that sustainability is well embedded into a company through Human Resources Practices, Marketing, Communication, Procurement, and CEO Leadership.

This book is for all those that want to effectively implement sustainable management and culture into their current business practices. The book runs simply through all of the steps to design and create a strategy according to the needs and expectations of all the stakeholders, ensuring everyone involved feels accountable and responsible for their piece of the pie.

It is also intended for all those studying or professionals that work as sustainability consultants or analysts to broaden their perceptions of sustainability implementation in a millennial digital world.

CHAPTER 1

SUSTAINABILITY: INTRODUCTION

Everything starts with posing the rights questions. For sustainability, we need to ask ourselves:

Who does what, when, and why? How does the company get the assets expected to endure and flourish? My emphasis here is on overseeing what I call the "physical measurements" of sustainable sustainability: the utilization of energy; the sources of energy, water, and materials; the decrease of waste and the environmental impact of a company's creation; and the effect of the utilization of whatever the company produces. The key inquiry to be tended: How does a company figure out how to lessen its negative effect on nature?

Sustainability effectiveness lies within the fate of the board. Sustainability must be coordinated into the core of what companies do. In the 20th century, companies focused on including universal activities, supply chains, data innovation, refined advertising, implementing procedures, ensuring money flow, bookkeeping, and new types of gathering forms.

Today, companies all through the globe are endeavoring to decrease their carbon footprint, limit atmosphere pollution, and sustainable hazard, upgrade strength, and incorporate sustainability into their tasks. The issue is they do not have the foggiest idea how to do it.

Organizations, governments, and charities are remembering they have to take these activities, yet their ability lies in their very own center capacities, not in creating sustainability designs and evaluating operational hazard inside this specific circumstance. To have important change on sustainability, companies, not-for-profits, and governments must carry it into the customary texture of authoritative setup.

Over the long haul, we ought not to require a "sustainability department;" each department should turn into a sustainability operating unit. Dealing with the physical components of sustainability should now be added to standard working methodology so all chiefs become sustainability supervisors. Huge thoughts, eager objectives, and delightful approaches do not convert into on-the-ground change until companies change.

The master plan issue we are trying to deliver at last comes down to the test of worldwide sustainability: how might we make a high throughput economy without decimating the planet? We are, actually, looking at sparing the planet for our kids.

The issues at the core of the sustainability issue are perplexing, worldwide, and fundamental and cannot be effectively unraveled. These difficulties require activity by partners at all levels. Be that as it may, making an interpretation of these thoughts energetically requires far more prominent hierarchical limit than we right now have. We may realize how to construct a LEED-certified energy productive structure, yet do we realize how to look after it? Frequently, we don't.

This hierarchical test is not generally recognized; it is not the flashiest issue that environment promoters are pointing out. In any case, companies are starting to comprehend that they cannot just make a sustainability department and anticipate that the issue should be unraveled. A significant number of our best heads realize they have to deal with the physical components of sustainability. There is ever more extensive acknowledgment of why sustainability is required, yet there is not yet a profound comprehension of how to do it. What we have to do is figure out how to fabricate that limit. As a scholastic, you will not be astounded to peruse that I accept the information we need must originate from scholastically thorough however applied research.

As an understudy of the board, I realize that every company and the authoritative condition is novel, yet there are examples and practices that can be distinguished and comprehended.

Under what conditions improve at lessening their utilization of energy and changing their wellspring of energy from high carbon to low? When do interests in shut frameworks building get made and when are they evaded? What sort of chiefs pays attention to environment components and what type believes it is a joke? In companies intensely subject to an all-inclusive production network, what techniques have succeeded and bombed when attempting to impact the conduct of providers?

As we gain more from the "bleeding edges" of sustainability the board, I will additionally be keen on the preparation and instruction needs that we distinguish. This will be added to our educational program in undergrad and graduate investigations in sustainability. Current courses should be adjusted and new courses should be created. I additionally anticipate that customary fields, for example, building, business, law, open organization, and general well-being, should add sustainability issues to their courses also.

There has been a propensity for some sustainable backers to concentrate on representative political fights, to some degree, as a method for picking up consideration and instructing people in general. The appreciation for imagery goes back to the beginning of advanced environment development.

The present unquestionably increasingly advanced variant of vehicle covering signals incorporate endeavors to strip benefits assets from petroleum derivative organizations and suing energy organizations for the environmental harm they have caused. It worked with the tobacco organizations; why not the oil organizations?

It may work in some sense, and it surely assumes a job in state-funded training. Be that as it may, the genuine work is persuading individuals to quit smoking and to utilize renewable wellsprings of energy. Strolling in the city of New York or urban communities anyplace on the planet, I see a lot of individuals smoking. The tobacco organizations appear to do OK.

Driving down those equivalent lanes is a lot of vehicles controlled by petroleum derivatives. We are altogether dependent on petroleum derivatives, and non-renewable energy source-based energy is integral to our day-by-day lives. I concur that putting resources into petroleum product organizations is an impractical notion, yet regardless I fill my tank with gas. Those organizations get our cash one way or the other.

The genuine work is getting an electric vehicle framework fabricated, propelling battery, and sustainable power source innovation, and actualizing these new advancements. I am keen on figuring out how to energize and speed up genuine change. Representative signals have a task to carry out; however, we ought not to mistake emblematic motions for operational reality.

Companies and social orders can be exceptionally delayed to change, however, after some time they do advance. In the previous century, a great part of the change has been in light of the improvement of new advancements. We have to adapt considerably more about our planet's environments and significantly more about the effect of our innovation on our planet. I accept that with insight, care, and new innovation, we can achieve a type of financial advancement that accomplishes both development and environment quality. In any case, it will be troublesome work, and it will not be accomplished rapidly.

CHAPTER 2

SUSTAINABLE DEVELOPMENT STRATEGIC TOPICS

CONTENTS

In 2019, preceded by political insecurity, there was an increase in environmental impacts/degradation and quickening paces of biodiversity misfortune. While the scale and multifaceted nature of the sustainability challenges we face is overwhelming, there is likewise an explanation behind good faith. We are seeing remarkable administration and advancement driven by urban areas, districts, and organizations, rising urban activism and promotion by Generation Z, and quickly developing mindfulness about issues, for example, plastic contamination. We anticipate that this developing force should drive truly necessary advancement on basic social and sustainable issues in 2019. Just because, we are additionally sharing further investigation of five areas: nourishment and agribusiness, innovation, pharmaceuticals, and social insurance, energy, and utilities and budgetary administrations. In our part inclines, we research the key issues confronting these quickly evolving enterprises and talk about rising arrangements.

Current endeavors stay deficient to change our course towards risky atmosphere change. A quickly warming atmosphere and inability to enough diminish ozone-depleting substance discharges the pressing requirement for a more prominent spotlight on flexibility measures. Quick, enormous scale interest in successful arrangements is expected to keep an unsustainable weather change from coming to conceivably cataclysmic levels.

2.1. RISING GLOBAL EMISSIONS

Falling costs for both sustainable power sources and flammable gas have brought about unobtrusive emissions decreases among top producers, for example, Europe and the United States.

In any case, none of the four biggest worldwide polluters (China, the United States, the European Union (EU), and India) are at present on track to do their part in worldwide endeavors to arrive at the Paris Agreement objectives.

Specialists report that worldwide ozone-depleting substance emissions rose by an expected 2.7% in 2018, arriving at a record-breaking high. This speaks to the second year of remarkable increments after a concise time of moderately stable emissions due to eased back monetary development.

Universal media center keeps on staying around outflows from power age, where the best progress is being made. Be that as it may, farming, and transportation are additionally significant supporters, and just restricted consideration is being given to these areas and other littler producers.

The most recent IPCC report featured the seriousness of effects we are now encountering at 1°C warming and expressed that 2°C of warming is currently considered profoundly hazardous.

Nations have consented to provide details regarding their emissions like clockwork beginning in 2024, as per the new Paris Agreement rulebook concurred at COP24 in December 2018 in Poland.

China proceeds to intensely put resources into sustainable power sources, electric vehicles (EV), and energy capacity, yet its absolute outflows are anticipated to keep moving until 2030. India's outflows are foreseen to top in 2033.

Air contamination identified with the utilization of non-renewable energy sources stays a significant worldwide issue. In excess of 3 billion individuals are breathing savage smoke in their homes from utilizing dirtying stoves and fills.

The EU has attempted to keep up outflows decrease energy, in enormous part because of Germany neglecting to change away from coal. In the meantime, the US carbon dioxide outflows rose by 3.4% in 2018, the greatest increment in 8 a long time.

"To have a chance of limiting warming to 1.5°C, greenhouse gas emissions must be cut in half by 2030 and be net zero by 2050." Diana Liverman, University of Arizona, lead author for the IPCC 1.5°C report (Figure 2.1).

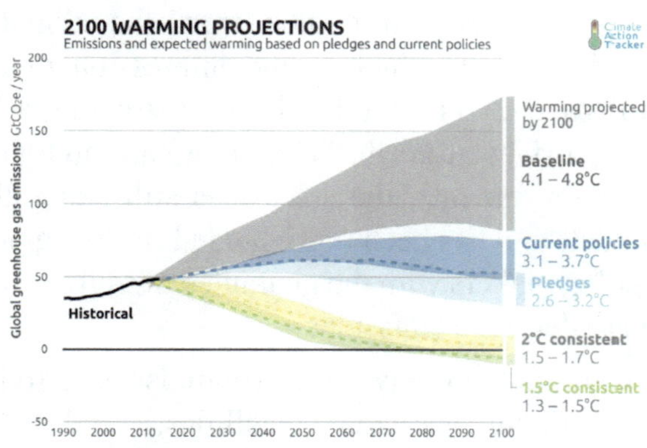

Figure 2.1: 2100 warming projections.

Source: Climate Action Tracker.

2.1.1. Adaptation Investment Is Urgent

The effects of 1°C warming are unmistakably more serious than anticipated. Extraordinary climate is causing billions in harm yearly, with flooding, storms, and dry season in Asia coupled with seething flames in the United States, Canada, Greece, and Sweden. While a few atmospheres relief and adjustment ventures are complimentary, as a rule, they are going after restricted government reserves.

Worldwide effects to urban situations are now evaluated at $314 billion every year and are anticipated to increment to more than $415 billion yearly by 2030.

Serious out-of-control fires in California caused billions of dollars in direct harm and protection payouts in 2018.

Warmth waves hit Europe in 2018, affecting financial execution and diminishing control age from wind turbines, featuring the dire requirement for additional interest in atmosphere flexible energy foundation and horticulture.

As per new gauges by FAO, dry seasons in Asia caused over $48 billion in lost harvests and domesticated animals somewhere in the range of 2005 and 2015.

A perilously warmed atmosphere is prompting recharged discourse among specialists about the upsides and downsides of geoengineering mediations. Geoengineering, for the most part, falls under two classifications: the expulsion and capacity of carbon from the climate and sunlight-based radiation the board, which is the impression of the sun's beams once again into space or on the other hand obstructing a little segment of daylight before it arrives at Earth. While there are numerous advocates of geoengineering arrangements, the science is still generally dependent on hypothetical models and has been reprimanded for being possibly risky and for exhibiting policymakers with the guarantee of a silver shot that may ease back endeavors to decrease emissions.

The New Carbon Economy Consortium is an activity of scholastic organizations, national research facilities, and Nongovernmental Organizations (NGOs) hoping to utilize rising advancements to change the economy by emphatically using carbon. The United States and Norway are at the front line of the drive for carbon catch what is more, stockpiling (CCS), yet so far, the European Commission (EC) has been tepid about this innovation.

Y Combinator, the tech hatchery celebrated for kick-beginning organizations like Airbnb and Dropbox, is currently emptying a huge number of dollars into financing geoengineering arrangements.

Geoscientists, environmentalists, and urban organizers are attempting to improve soil to retain higher measures of carbon dioxide in both farming and urban situations. Other incipient innovations getting subsidizing incorporate hereditarily designed phytoplankton, direct air catch, marine cloud lighting up, and electrochemistry.

Regardless of expanding enthusiasm for geoengineering arrangements, a developing number of specialists are cautioning about potential risks related to the innovation, counting obscure and conceivably disastrous reactions, significant expenses, and the fracture of worldwide endeavors on ozone harming substance decrease.

"Geoengineering is the brave new world of climate mitigation. While the scientific jury is still out on risk and effectiveness, what's already becoming clear from our initial research is that any silver-bullet climate solution brings with it an additional 'moral hazard' that it may desensitize the general public to climate risk and sap their motivation to support decisive mitigation." Victoria Campbell-Árvai, a Research Scientist at the University of Michigan's School for Environment and Sustainability (Figure 2.2).

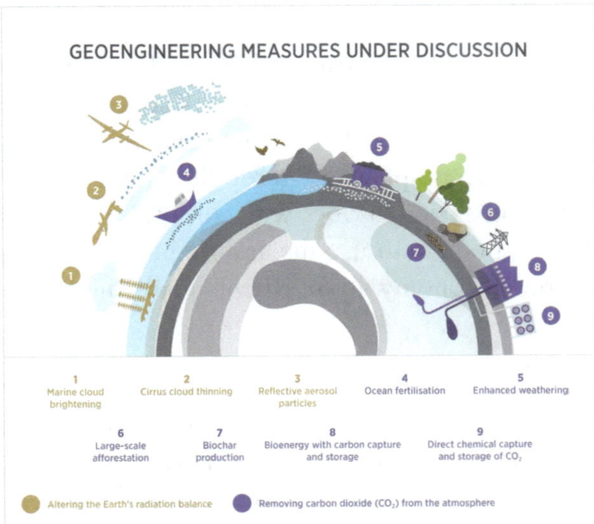

Figure 2.2: Geoengineering measures under discussion.

Source: IASS.

2.1.2. Future Expectations

We hope to see billions of dollars of extra harm and a lot more lives lost because of extraordinary climate occasions. Versatility, adjustment, and obligation with respect to framework that is badly arranged for progressively outrageous climate will be front what is more, place for the two governments and business. Organizations will quicken activity to build atmosphere readiness in both direct tasks and their inventory chains. Geoengineering advancements could see more prominent speculation from the private area furthermore, restored exchange at the strategy level.

1. **Organize Interest in Atmosphere Strength and Adjustment:** Organizations need to build the versatility of their stockpile chains and direct tasks. This will require interest in atmosphere hazard evaluations and situation arranging.
2. **Shape Strategy and Campaign Governments for More Atmosphere Activity:** Organizations need to have a progressively dynamic influence in national and worldwide endeavors to execute the Paris Agreement and apply more noteworthy weight on governments to restrict a worldwide temperature alteration. This implies supporting for activities like a cost on carbon, forceful arrangement of renewables and EV, and atmosphere brilliant farming arrangements.
3. **Increment Collaboration Efforts to Scale Low Carbon Economy Solutions:** Quickly decreasing ozone harming substance outflows and expanding the versatility of framework may be accomplished through multi-area joint effort. Organizations need to work with corporate cohorts, governments, NGOs, and others to use the assorted aptitudes required scaling low carbon economy arrangements in all enterprises.

2.2. CITIZEN POWER

Citizens demand a more sustainable, equitable world from companies and institutions. Engaged citizens, joined by an increasing number of Gen Z teens, are taking to social media and the streets to demand action from governments and businesses on pressing social and environmental challenges.

Savvy companies are tapping into shifting consumer values and empowering people to act on important issues like climate change.

2.2.1. Power to the People

Around the world, a small but growing number of citizens are rising up to challenge institutions, companies, and governments to deliver a more equitable and sustainable society. They are increasingly using social media and other digital technologies to spread the message, elevate important issues, and mobilize people.

Civil disobedience movement Extinction Rebellion—a UK-born action group that calls for mass economic disruption using non-violent direct action and civil disobedience to prevent catastrophic climate change—has now spread to more than 35 countries around the world. Americans took to the streets in 2018. Millions protested against gun violence, gender inequity and the separation of children from their migrant and asylum-seeking parents.

One in five Americans have protested in the streets or participated in political rallies over the last 2 years. Of those, 19% said they had never before joined a protest march or a political gathering.

In one of the biggest movements for women's rights in India, 5 million women lined up across the length of the southern state of Kerala to demand an end to outdated social norms that prevent women of childbearing age from entering places of worship.

In France, thousands of working-class protestors known as the gilets Jaunes (yellow vests) have taken to the streets to protest the newly-introduced green taxon diesel—highlighting the need for a more just energy transition—and call for an increase in the minimum wage.

Employees at multinational firms are increasingly making their voices heard, with Amazon employees protesting the sale of facial recognition software to law enforcement and unsafe work practices.

Meanwhile, over 1000 Google employees signed letters protesting the company's involvement with AI military development for the Pentagon, as well as its decision to aid China's online censorship. "Empowering Action at Home": A study of 30,000 people in 35 countries found that 62% of customers want companies to take a stand on issues such as sustainability, transparency, and fair employment practices."

"A majority of global consumers are concerned about climate change, but only a handful of these feel empowered to take action in their own lives. Global companies, like IKEA, have an important role to play in inspiring and supporting more people to take action." Pia Heidenmark Cook, Chief Sustainability Officer, IKEA Group.

2.2.2. Gen Z Coming of Age

Conceived when the new century rolled over, Gen Z is as of now entering the workforce and will overwhelm recent college grads to turn into the biggest age of purchasers by 2020. While many are still too youthful to even consider voting, Gen Z is now significant influencers, locks in organizations and governments on a scope of social and environmental issues.

Youth turned marching through Main Street for COP24, the UN Climate Change Conference in Poland, and sent a record-breaking message calling for worldwide pioneers to meet the 1.5°C objective as 125,000 postcards spread out on Aletsch ice sheet in Switzerland.

In Australia, a large number of youngsters from in excess of 20 urban communities left their homerooms to fight the Australian government's inaction on environmental change as a major aspect of the Strike 4 Climate Action, roused by 15-year-old Swedish understudy Greta Thunberg.

The understudy drove March for Our Lives saw in excess of a million Americans take to the avenues supporting for more grounded weapon guidelines.

As acts of mass violence proceed over the US, American secondary school understudies are putting pressure on financial specialists, for example, Vanguard and BlackRock to drop firearm producers from their speculation portfolios.

An ongoing report found that 68% of youngsters matured 14–18 in the UK have taken an interest in volunteering or different types of social activity.

A multi-year-old young lady reached Warriors basketballer Steph Curry on Twitter to inquire why his Under Armor shoes were accessible for young men, and Curry and Under Protective layer helped the issue in under 24 hours.

2.2.3. Empowering Action at Home

Individuals progressively need organizations to make significant move on environment and social issues. Driving organizations are discovering approaches to empower individuals to make simpler, sustainable, practical changes to their lives.

An investigation of 30,000 individuals in 35 nations found that 62% of clients need organizations to stand firm on issues, for example, sustainability, straightforwardness, and reasonable work rehearses.

Research by IKEA and GlobeScan found that a lion's share of shoppers needs to lessen their carbon outflows, however many are reluctant to pick less advantageous or, on the other hand, progressively costly choices.

Organizations including Patagonia and IKEA are implanting manageable advancement into their plans of action with the objective of engaging residents to lead more sustainable lives.

Unilever backup Ben and Jerry's is using blockchain to engage clients to carbon counterbalance each individual scoop of frozen yogurt.

Worldwide office-sharing organization WeWork has gone sans meat at its 502 areas—with meat never again served at staff occasions and WeWork workers not ready to cost meat dinners—in an offer to diminish carbon discharges.

2.2.4. Future Expectations

Resident drove activism will keep the weight on governments and organizations to address squeezing social and environment issues. Desires for organizations to address issues, for example, environmental change, single-use plastics, and gender equality value will likewise keep on developing. Clients, particularly youth, will progressively move their loyalties to organizations they see to convey social and sustainable incentives to society.

1. **Convey an Incentive to Society**: In 2019, organizations need to reassess the sustainable and social effect of their center products and administrations and move their plans of action towards producing long term an incentive for society on the off chance that they are to stay aware of moving customer desires.
2. **Increment Straightforwardness**: So as to construct trust and reliability in 2019, organizations should grasp progressively extreme types of straightforwardness. Blockchain and web-based social networking will be fundamental apparatuses.
3. **Engage Purchasers**: Driving organizations ought to put resources into building brand control by empowering customers to make manageable way of life changes that are both helpful and cost-aggressive.

2.3. SECURITY THREATS

Exchange wars, cybersecurity breaks, and environmental change undermine worldwide security and financial soundness.

The worldwide network is confronting difficulties on universal exchange, the effects of environmental change and the developing seriousness of digital fighting, which are all affecting advancement on reasonable improvement.

Governments and worldwide organizations are progressively powerless against cybersecurity breaks, debilitating their companies just as resident trust. Exchange pressures undermine a debilitating worldwide economy, while environmental change and different variables heighten clashes in the Middle East, Central America, and Africa and uproot millions.

2.3.1. The Cyber Battleground

From state-supported cyber-attacks, to psychological militant cells, and the developing advancement of solitary programmers, the dangers of cyberwarfare are developing. Expanded digitization and mechanization, particularly IoT and distributed computing, make greater open door for cyberattacks. Hacking development keeps on outpacing the security incorporated with existing innovation. Government guidelines fall insufficiently behind.

The World Economic Forum's Global Risks Report 2018 recognized cybersecurity breaks as probably the most serious hazard to mankind. In its Top 10 dangers digital assaults are positioned third as far as Likelihood.

New deep fake innovation permits the production of carefully controlled recordings that can approach flawlessly copy anybody, from heads of state to military commanders.

Malevolent chatbots are getting progressively far reaching. Assaults presently start with fundamental content-based bots, however, could begin to utilize human discourse bots to capture exploited people over the telephone.

The year 2018 saw various prominent digital assaults hit enormous organizations including Facebook, Google+, British Airways, Marriott, and T Mobile. WhatsApp, a mainstream auxiliary of Facebook, was utilized to spread disinformation by Russian programmers in the number one spot up to the 2018 Brazilian political race and may have assumed a job in the appointment of Jair Bolsonaro.

Confronting analysis over the abuse of its foundation during national decisions in Brazil, the US, France, and the UK, Facebook has declared designs to increment advertisement straightforwardness and safeguard against remote resistance in front of the 2019 Lok Sabha surveys in India.

"Countries need to work together to tackle challenges that extend beyond their own borders. Countries should cooperate to reduce trade costs further and resolve disagreements without raising distortionary barriers. Cooperative efforts are also essential for enhancing cybersecurity, tackling corruption, and mitigating and coping with climate change." International Monetary Fund, World Economic Outlook 2018 (Figure 2.3).

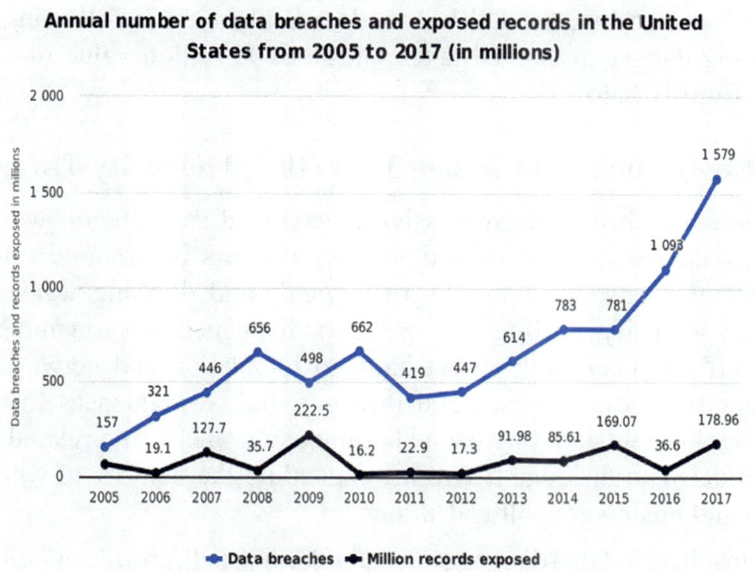

Figure 2.3: Data breaches.

Source: Identity Theft Resource Center.

2.3.2. Exchange Tensions Threaten Global Economy

The Trump administration is endeavoring to decrease imports from nations with which the US has an exchange shortage, upsetting exchange arrangements with the EU, Canada, Mexico, South Korea, Japan, and, above all, China. Progressing pressure between the world's two biggest economies over exchange is expanding the probability of a worldwide financial downturn by 2020.

The breakdown of two-sided exchange talks and the acceleration of duties and other protectionist resistances by the US and China is starting to hurt the two economies.

Exchange pressures could decrease universal development by around 0.5% by 2020, the IMF reports.

The CFO of Chinese telecoms organization Huawei was captured in Canada over potential infringement of US endorses on Iran, shocking China, and confusing exchange arrangements between the world's two biggest economies.

As of composing, the US has made a deal to avoid following up on its ongoing danger to expand taxes on the $200 billion value of Chinese imports from 10% to 25%.

2.3.3. Environmental Change Tops Global Security Threats

Environmental change is progressively perceived as the main worldwide security risk. Atmosphere-related security dangers incorporate effects to fundamental assets, for example, rural yields and drinking water, which can thusly intensify existing clashes, as has happened in Syria and Yemen. Further effects incorporate direct harm to foundation and networks from out-of-control fires, tempests, and flooding, and consequences for human well-being brought about by ozone harming substances, and related toxins. These effects can dislodge networks, expanding the number of worldwide outcasts and raising geopolitical strains.

Atmosphere-related dangers, for example, dry spell, crop disappointment, typhoons, and rapidly spreading fires, are huge supporters of contention and dislodging of huge quantities of individuals in Syria, Yemen, Kenya, and Sudan. The most recent National Climate Assessment directed by the US government presumes that environmental change is "exhibiting developing difficulties to human well-being what's more, security, personal satisfaction, and the pace of monetary development" and accordingly presents noteworthy national security dangers.

Right now, more than 65 million individuals are uprooted internationally for some reasons counting struggle, starvation, and catastrophic events, which is the biggest number ever recorded by the UN Refugee Agency. The World Bank predicts that over 140 million individuals will be dislodged by environmental change alone by 2050.

2.3.4. Future Expectations

The worldwide network will see much more prominent union of geopolitical and reasonable improvement motivation, with declining financial, physical, and computerized dangers in 2019. Environmental change impacts on worldwide security will turn out to be more conspicuous. The expanding number of cybersecurity breaks and exchange clashes are probably going to compound geopolitical pressures and cost the worldwide network billions of dollars. A potential worldwide financial downturn may slow advance on sustainability in the event that organizations decide to react by decreasing their sustainability spending plans:

1. **Put Resources into Cybersecurity:** The two governments and businesses are neglecting to stay aware of programmers. All organizations need to expand interest in cybersecurity, and tech organizations specifically need to attempt to work all the more intimately with governments to assist them with sanctioning guidelines to all the more likely protect the two organizations furthermore, residents.
2. **Draw in Governments on Exchange**: Enormous organizations need to cooperate, and with exchange relationships, to utilize their impact with governments and multilateral companies to endeavor to balance out the present unstable exchange relations between the US and other huge economies, most strikingly China.
3. **Dissect Geopolitical Dangers:** Organizations that dissect geopolitical patterns will have the option to more readily envision, what is more, be readied, for interruption regardless of whether this comes in progressively capricious and entangled ways.

2.4. SAVING ECOSYSTEMS

Energy develops to handle plastic waste as biological systems are confronting phenomenal dangers. Our property and marine environments are confronting phenomenal dangers. Species are vanishing at disturbing rates, land corruption is undermining the prosperity of a large number of individuals, and compounding air contamination is costing a huge number of lives every year, and plastic contamination in seas and ashore is arriving at perilous levels.

China's choice to boycott plastic and e-waste imports has tossed waste and reusing markets into unrest, uncovering the greatness of plastic contamination, and raising the criticalness to lessen and reuse unmistakably more materials.

2.4.1. Biodiversity at Breaking Point

Appraisals of the Earth's biological system present a depressing picture. While mindfulness about biodiversity misfortune is developing, consideration paid by organizations and different companies to the issue stays restricted. As per the most exhaustive evaluation of worldwide biological system well-being in late decades, biodiversity is declining at a disturbing rate crosswise over a large portion of the characteristic world. Species are vanishing multiple times quicker than their common rate of termination. In its Living Planet Report, WWF finds "detonating human utilization" was the primary purpose for a monstrous drop in worldwide sustainable life populace in later decades. Somewhere in the range of 1970 and 2014, misfortunes in vertebrate species—warm-blooded animals, fish, and birds of land and water, and reptiles—found the middle value of 60%. Air contamination levels are arriving at basic levels. WHO evaluates that 9 out of 10 individuals inhale air containing significant levels of toxins, and around 7 million individuals bite the dust each year from presentation to contaminated air. In excess of 75% of land, zones are currently considerably debased, which adversely impacts the prosperity of more than 3.2 billion individuals. Sustainability specialists accept that the worldwide network has made constrained progress on SDG 14: Life below Water and SDG 15: Life on Land. They are among the sustainable development goals (SDGs) that get minimal consideration from business and different companies.

"Earth is losing biodiversity at a rate seen only during mass extinctions" Living Planet Report, WWF (Figure 2.4).

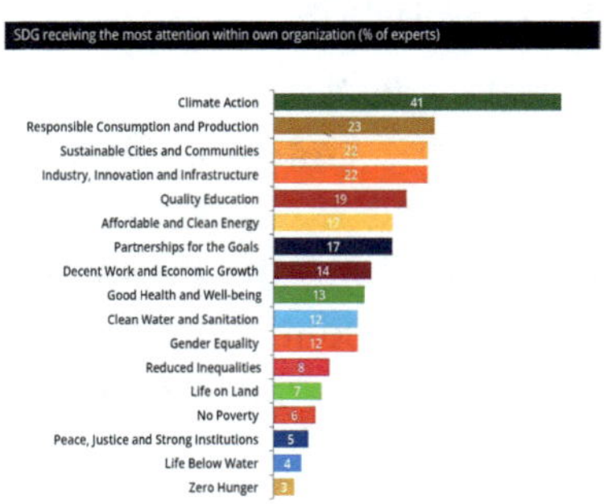

Figure 2.4: SDG 14: Life below water and SDG 15: Life on land is currently among sustainable development goals that receive the least attention by businesses and other organizations.

Source: Globe Scan-Sustainability Survey.

2.4.2. Waste and Plastic Hit Headlines

Worldwide mindfulness about plastic contamination and its effect on our biological systems proceeds to develop. A few significant worldwide alliances have been declared to handle the issue. Organizations are encountering developing weight from customers to diminish plastic waste.

Before reporting a boycott in January 2018, China imported more than 66% of the world's plastic waste. Nestlé is working towards plastic waste issue by making an Institute of innovative science research, which will concentrate on improving recyclable, biodegradable, and compostable packaging.

The New Plastics Economy Global Commitment unites organizations, government, and common society behind a dream for accomplishing a round economy for plastic. The 290 companies that have marked the responsibility speak to 20% of all plastic packaging delivered all around.

UNEP's Global Plastics Platform means to drive cross-fringe joint effort and learning among nations and urban communities.

Marine plastic issues are increasing expanding unmistakable quality in Asia. Course Capital what is more, Ocean Conservancy reported at any rate \$150 million of venture capital for waste and reusing advancements that keep plastic waste from entering the sea from Asian nations.

WWF reported a 3-year, \$7.5 million No More Plastics in our Ocean activity supported by the Norwegian government to propel a worldwide administration answer for marine plastic contamination (Figure 2.5).

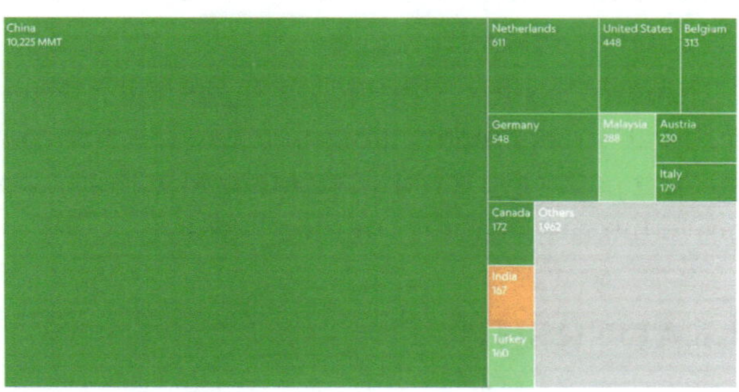

Figure 2.5: Plastic waste imports, millions of metric tons, 2016. For many years, China was the main importer of plastic waste, which changed in 2018 when it announced a ban.

Source: National Geographic.

2.4.3. Future Expectations

Worldwide mindfulness about waste and plastic contamination will keep on developing. A number of coordinated efforts have just been reported however a lot more prominent exertion will be required to handle the issue. China's choice to boycott wastes imports will proceed to put pressure on governments and organizations to search for fundamental arrangements on a worldwide scale. While mindfulness about biodiversity misfortune contrasted with other practical improvement issues has been low, a developing number of NGOs are attempting to transform it. Open and private companies the same will confront developing desires to have an increasingly dynamic influence in saving area and marine biological systems.

1. **Address Plastic Waste Over the Whole Value Chain:** Organizations crosswise over numerous ventures—not just the ones legitimately capable for the greater part of plastic waste (nourishment and agribusiness, retail, and so on.) yet additionally those less related with plastic contamination (innovation, pharma, and so forth.)—will confront developing pressure by governments, clients, and different partners to address plastic waste. Organizations need to research their whole value chain to execute arrangements that would diminish waste and plastic contamination.
2. **Biodiversity as a Material Issue:** Not many organizations distinguish biodiversity as a material issue, yet as mindfulness about land and marine environments emergencies develops, organizations will be relied upon to research their effect—and reliance on biodiversity.
3. **Put Resources into Round Arrangements:** Putting resources into round arrangements along the whole value chain gives one of the best pathways for organizations to address waste issues and improve the sustainability of their tasks.

2.5. NEW LEADERSHIP

Urban areas, locales, and businesses keep on sloping up authority as trust in national governments struggles. Worldwide governments are bearing a continuous emergency of trust. Non-state entertainers, for example, city, and provincial governments and the business network, are venturing into the authority void and making a move to make a progressively manageable, impartial future.

2.5.1. Doubted Governments

Crosswise over significant European powers, the US, Canada, Russia, Australia, and Latin America, trust in governments stays low. This has been filled by expanding political polarization and deficient reactions to issues, for example, disparity, and environmental change.

Trust in national governments in a lion's share of European nations stays low, counting the UK, Germany, France, Spain, Italy, and Russia.

Latin America likewise has low degrees of trust in government with Brazil, Argentina what is more, Colombia all scoring under 50%.

The US Administration is bearing an immense emergency of trust—the first occasion when that such a huge decrease in trust has happened outside of a monetary downturn or national calamity as indicated by Edelman Trust Barometer.

In China, state-controlled assessments of public sentiment demonstrate high trust in government, yet this information has been raised doubt about by specialists who express that the administration's dictator control has brought about expanding doubt and loss of social capital.

Specialists evaluated national governments as having the least fortunate performance on tending to sustainable advancement out of any worldwide organization.

2.5.2. Urban Communities and Regions Lead

With worldwide pioneers neglecting to make a satisfactory administrative move on a scope of issues—counting atmosphere alleviation and adjustment, sea plastics, waste, and well-being—urban areas, what is more, provincial governments are assuming the job of sustainable advancement pioneers.

In the US, trust in neighborhood governments is at its most elevated since 2008. California leads subnational endeavors to control environmental change, with objectives to move to 100% sans carbon power by 2045 and to become carbon impartial by the same year.

Around 27 worldwide urban communities, including London, New York, Sydney, and Rome, have as of now arrived at top outflows and have effectively diminished their emissions for the most recent 5 years while keeping up powerful monetary development.

San Francisco now occupies 80% of its loss from landfills. It additionally joins New York, London, Dubai, and 20 other worldwide urban areas in having an objective a zero-waste objective.

Company for Healthy Cities is another coordinated effort between a worldwide system of urban communities, WHO, and Bloomberg Philanthropies to spare lives by anticipating noncommunicable ailments and wounds.

In excess of 20 worldwide urban communities including London, Los Angeles, Paris, Mexico City, Seoul, and Tokyo have promised to guarantee a significant region of every city is zero emissions by 2030 and to buy 100% zero-discharge transports from 2025 onwards.

Ukrainian urban areas have built up an objective of moving to 100% renewable energy by 2050, in an offer to address environmental change and increment energy autonomy.

2.5.3. Business Leadership Is Increasing

Business is progressively venturing into the void left by worldwide governments, setting aggressive new social and sustainable objectives, and participating in radical straightforwardness endeavors that help to fabricate trust and believability with the two shoppers and workers.

In excess of 500 worldwide organizations are making a move on science-based atmosphere targets. Innovation organizations including Salesforce, Autodesk, Lyft, and Arm have framed the step-up coalition with the objective of using rising advances to lessen ozone-depleting substance emissions overall industry areas.

Driving Japanese insurance agency Sompo Japan Nipponkoa has set SDG targets for each territory of its business, including an objective to create products and administrations identified with disaster avoidance and alleviation.

Around 150 organizations have focused on 100% sustainable power source through RE100. What is more, 23 multinationals have resolved to zero outflow vehicles through EV100.

Google as of late met its 100% sustainable power source objective, joining Apple and Microsoft.

Illustrious Dutch Shell has declared that it will interface official compensation to its carbon decrease objectives.

"With several governments dialing back from global commitments and reducing environmental protection, we need- and are seeing-the non-state actors step up to the challenge. We need a New Deal for nature and people." Kavita Prakash-Mani, WWF (Figure 2.6).

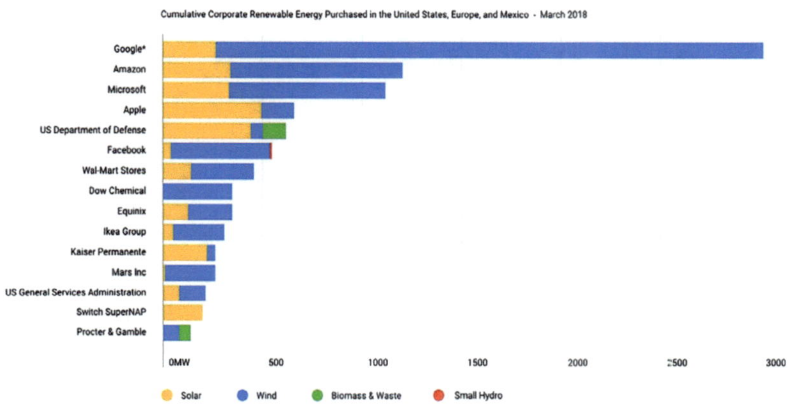

Figure 2.6: Royal Dutch shell executive remuneration and carbon reduction goals.

Source: Royal Dutch Shell.

We hope to see proceeded with low degrees of trust in national governments and fortifying authority and aspiration of urban communities, provincial governments, and organizations. Proceeding with vulnerability in Europe due to Brexit and expanding polarization in the US is probably going to additionally hose trust in the political frameworks.

Buyers will be expanding hoping to organizations to act in agreement to their qualities and all the more effectively advocate on social and environmental issues.

Set intense objectives: Business should keep on inclining up administration and aspiration on conveying social and environmental advantages to society by setting and providing details regarding execution against intense sustainability objectives.

Increment straightforwardness: One of the best approaches to build trust is making information—on everything from decent variety to official compensation and carbon discharges—open and available to the general population.

CHAPTER 3

STRATEGIC TRENDS PER SECTOR

CONTENTS

3.1. FOOD AND AGRICULTURE

"Demand for food poses one of the biggest dangers to our planet. It's the leading cause of deforestation, destroying countless habitats and threatening wildlife to the point of extinction."—Tanya Steele, WWF UK Chief Executive.

There is a critical need to make our nourishment framework unquestionably progressively feasible. Developing populace and related nourishment utilization is fueling biodiversity misfortune on an extraordinary scale, with 40% of the world's once-forested land now cleared for human horticulture and settlements.

Biological systems ashore and in the sea are being crushed by plastic packaging—an enormous level of which originates from nourishment and drink items. Entertainers in the nourishment esteem chain, from cultivators to retailers, are giving expanding consideration to both nourishment furthermore, plastic waste—investigating ways packaging can be decreased without settling item time span of usability or sanitation.

3.1.1. Adjusting Packaging and Food Waste

Plastic packaging is progressively in the media spotlight, with shoppers requesting activity from organizations on single-use plastics. Food industry producers are feeling the strain to diminish both plastic packaging, just as food waste. Research shows that shrewd and economical packaging can help decrease nourishment waste, and a developing number of organizations are putting resources into new innovation arrangements.

The worldwide expense of nourishment waste is assessed to be nearly $1 trillion per year, with up to 25% of private waste emerging from deficient packaging.

While some accept single-utilize plastic packaging has prompted an expansion in the measure of nourishment waste, many contend that without plastic packaging, the expense of nourishment waste could rise.

There are a developing number of organizations putting resources into maintainable packaging. Kroger has collaborated with Apeel Sciences that makes a defensive film from rural side-effects and waste, and Florida-based SaltWater Brewery is utilizing a biodegradable and compostable six-pack ring configuration produced using wheat and grain.

UK retailers, Tesco, and Waitrose, have propelled new supportable packaging to expand the time span of usability of avocados and Morrisons are because of preliminary a film packaging arrangement which can stretch out timeframe of realistic usability by dependent upon two to four days.

New businesses in the US handling the test of nourishment waste got over $125 million in funding and private value interest in 2018.

"An issue of packaging is an issue of food waste, and an issue of food waste is an issue of hunger and food availability—we need to find and focus on the balance."—Jackie Suggitt, Director of Business Initiatives, ReFED.

3.1.2. Singling Out the Kids

The pace of corpulence in kids and teenagers has risen ten times comprehensively in the previous 40 a long time, with nourishment advertising being one of the most perceived givers. Kids in created nations keep on being raised in a stoutness advancing condition, encompassed by cheap and influentially showcased fatty nourishment. Youngsters are presented to twice the same number of low-supplement nourishment as sound nourishment advertisements; with examine exhibiting that such presentation impacts affect kids' weight control plans.

WHO has discovered that current strategies and guidelines focused on nourishment advertising to youngsters in many nations are "especially lacking," with kids still uncovered to unsafe business messages advancing nourishments high in fats, salt, and sugar.

Weight specialists are thinking about the case against the nourishment business in the light of developing examination recommending that an ascent in lousy nourishment neuromarketing that could commandeer a youngster's cerebrum.

Of the evaluated £296.6 million spent on nourishment showcasing in the UK every year, just 5% is allotted to leafy foods.

The Peas Please activity is proposing another Veg Ad Fund to guarantee vegetables get comparable degrees of showcasing speculation as marked chocolates, cheap food outlets, and soda pops. The crusade will dispatch in 2019 and has support from significant general store chains.

McDonald's as of late reported goals to consent to stricter wholesome rules and capable advertising, utilizing inventive promoting to help serve more organic product, vegetables, entire grains, lean protein, and water in Happy Meals.

3.1.3. Dietary Diversity

Biodiversity is basic to guaranteeing sound biological systems; nourishment security, and sustenance. Be that as it may, nourishment frameworks are a key driver of biodiversity misfortune around the world. In spite of nearly 7,000 crops being consumable, most of the world's worldwide energy admission originates from just three: rice, wheat, and maize. The subsequent monoculture, combined with expanded pressure on agrarian land for domesticated animals cultivating, is fueling biodiversity misfortune on an uncommon scale.

Living Planet Report by WWF recognized human utilization to be behind an uncommon loss of vertebrate species.

Somewhere in the range of 2010 and 2050, the natural impacts of the nourishment framework could increment by 50–90%, coming to past planetary limits if no activity is taken.

Shoppers are progressively mindful of the effect of diet on biodiversity. One out of eight Britons is presently veggie lover or vegetarian and a further 21% guarantee to be flexitarian.

India has for some time been a worldwide pioneer, with roughly 20% of the populace being veggie-lover, yet progress is likewise being made in China, where veggie lover eateries are starting to multiply in significant urban areas, and meat deals are starting to decrease.

Retailers are urging customers to receive progressively manageable eating regimen. Tesco is collaborating with WWF to urge customers to purchase moderate manageable nourishment.

Walmart and Unilever are teaming up with neighborhood governments to stop the loss of biodiversity in their sourcing locales.

3.1.4. Feed Behind Our Food

Animal feed is an imperative, yet inconspicuous, contribution to the nourishment business that has noteworthy suggestions for ecological wellbeing and nourishment security. Horticulture is mindful for 14.5% of every single outflows, with feed generation bookkeeping for 45% of this aggregate. In numerous nations domesticated animals generation is quickening deforestation, biodiversity misfortune, and water shortage.

While the utilization of most nourishment waste in feed is illicit in the EU, it is advanced in other portions of the world. In Japan and South Korea, around 40% of nourishment waste is reused as animal feed.

Weight from promotion gatherings and even pig ranchers to relegalize heat-treated swill in the EU is expanding, with explore proposing this would altogether decrease land utilization.

Organizations, for example, Calysta, Protix, Ynsect, and Evonik are putting millions in creative recognizable feed fixings. This incorporates bug-based protein, oil from characteristic marine green growth, feed added substances like amino acids, and protein got from methane-eating microscopic organisms.

Pet nourishment speaks to as much as 30% of all meat utilization in the US, furthermore, there is a developing requirement for maintainable other options. Wild Earth is as of now, the greatest player, making pooch and feline treats from parasite and lab-developed mouse meat (Figure 3.1).

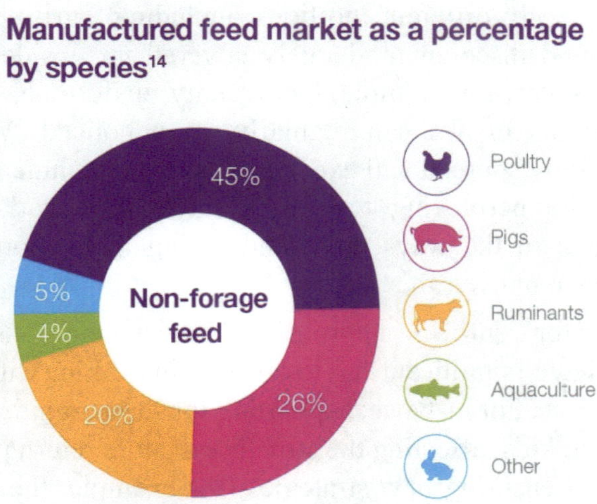

Figure 3.1: Manufactured feed market as percentage of species.

Source: International Feed industry.

3.1.5. Future Expectations

Issues identified with waste and packaging will keep on drawing in examination in the coming years, putting pressure on organizations and other key partners to search for fundamental arrangements. Poor industry principles in regards to showcasing of low nourishment, unhealthy nourishments, and refreshments to youngsters will likewise create progressively escalated discussions—with an expanding number of organizations making the willful move to raise industry guidelines. Securing biodiversity, health, and plant-based weight control plans will likewise be a bigger concentration for promoters, business, and customers.

Actualize circular arrangements: Producers to retailers need to grasp round intuition to address current foundational impediments. This methodology looks past setting eager reusing focuses for packaging and contemplates endeavors to upcycle nourishment waste into new items—where waste items from one piece of the framework become contributions to another.

Improve perceivability of concealed issues: The nourishment security banter has to a great extent concentrated on the homestead part and on exchange, with different portions including agri-inputs, handling, coordination, and discount moderately covered up. Inside these regions, sustainability issues, for example, biodiversity misfortune, water shortage, and human rights breaks can regularly go unnoticed. With customers requesting morally sourced and excellent items, nourishment organizations must handle the perplexing test of production network perceivability and should look to the ways that trend-setting innovation, for example, blockchain, can empower this.

Promoter for shrewd approach: Nourishment and refreshment organizations have a significant task to carry out in working with policymakers to catalyze change. For instance: supporting for stricter guidelines in regards to publicizing to kids; affecting the structure of store return plans for plastic assortment; and ensuring new strategies (for example, the US Farm Bill) consolidate activity with respect to the squeezing need to improve soil wellbeing.

3.2. TECHNOLOGY

"In 2019, there will be increasing focus on opportunities to apply technology to help a range of industries drive smarter decisions and resource efficiency, develops circular economy solutions, and drive improvements in supply chain responsibility practices."—Suzanne Fallender, Director, Corporate Responsibility Office, Intel Corporation.

The tech division needs to grasp straightforwardness what's more, city duty to satisfy its latent capacity. Cultural trust in the innovation division is declining—driven by significant information protection ruptures, charge evasion, the utilization of innovation stages by rebel players to undermine vote-based procedures and an absence of sufficient government guidelines, what's more, oversight.

However, against this background of doubt, tech organizations have kept on pushing forward with developments that drive financial development and assist humankind with handling squeezing social, what's more, ecological difficulties. The area remains at an urgent minute where changes will be important to modify trust and understand the part's maximum capacity to move society toward a progressively reasonable future.

3.2.1. Moral Code under Fire

For long the most confided in part, trust in innovation organizations and their administrations (particularly enormous Internet organizations) has been declining, showing reputational and business dangers for the segment. Innovation organizations are progressively tested by customers and different partners to exhibit the positive social effects of their items and administrations on the networks in which they work. Rising innovations, for example, self-ruling vehicles, blockchain, and AI remain the most inadequately comprehended furthermore, least trusted by people in general.

Trust in web search tools and internet-based life stages has diminished in numerous nations, with the steepest decrease in the US Tech organizations' security rehearses, especially Apple, Google, and Amazon have been the fundamental driver of declining trust. A review directed by Fortune magazine seen Facebook as the least trusted of significant tech organizations with regards to shielding client information.

Edelman's 2018 Trust in Technology report found that rising advancements are the least trusted by the overall population with blockchain trusted by marginally under 50%, self-driving vehicles at 50% and man-made consciousness at 56%.

Around half of US grown-ups might want innovation organizations to be all the more intensely controlled, as per an overview by Pew Research Center.

Americans are erasing or changing the manner in which they use Facebook, and a few brands (P&G and Mars) are slicing computerized advertisement spends because of information wellbeing concerns.

3.2.2. Simulated Intelligence for Good

Artificial Intelligence (AI) is set to change businesses over each area and can possibly quicken endeavors to handle the world's greatest cultural and ecological difficulties. The test for tech organizations will be to guarantee innovative work constructs frameworks which have value, security, and morals at their center.

Accenture gages that AI could twofold financial development rates in 2035 by changing the idea of work and could expand work profitability by up to 40%.

Studies gage that somewhere in the range of 14 to 54% of US laborers have a high likelihood of seeing their occupations mechanized throughout the following 20 years.

The yearly estimation of AI is required to be most elevated in the gadgets, customer bundled products, and retail parts, as indicated by the McKinsey examination. Overall spending on subjective and man-made brainpower frameworks is relied upon to reach $57.6 billion out of 2021.

Organizations are taking activities to demonstrate that AI advances can be utilized for good –, for example, Microsoft AI for Earth which intends to make AI-based arrangements over farming, water, biodiversity, and environmental change.

Studies have featured how AI can fortify race and sex inclinations due to innovation being created by groups overwhelmed by white men (Figure 3.2).

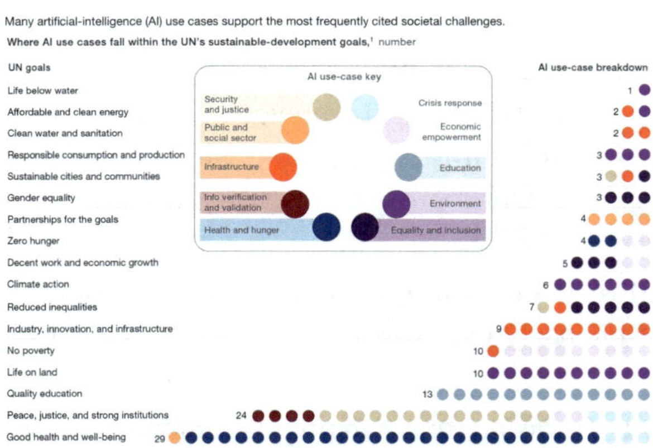

Figure 3.2: AI potential.

Source: McKinsey Global Institute.

3.2.3. Battling Waste with Reuse

Before it restricted e-waste imports, China was taking in 70% of the world's electronic waste. Since the boycott, Thailand, Laos, and Cambodia have gotten top goals for e-waste from North America and the EU. Therefore, these Asian nations are currently confronting basic degrees of natural contamination from e-waste reusing. Looked with developing worry about waste and store network impacts, a developing number of innovation organizations are receiving inventive round economy arrangements so as to diminish dependence on limited materials, recoup important minerals and arrive at new clients and markets.

Mac has uncovered its new MacBooks made with 100% reused aluminum. The organization as of late reported an aspiring objective of moving to a shut circle store network where items are constructed utilizing just sustainable assets or reused materials.

Dell has been an industry head-on reusing and has as of late helped discovered the next wave plastics alliance, which expects to make a worldwide system of supply chains intended for gathering and reusing plastics.

An ongoing report found that numerous buyer items produced using reused electronic waste contain harmful synthetics and approaches the EU to characterize e-waste as dangerous.

Innovation organizations are confronting developing reaction for arranged outdated nature of their items, and right to repair bills are being set up for 2019 crosswise over numerous US states as force develops to slow the measure of new equipment available.

3.2.4. Future Expectations

2019 will probably be another unstable year for the tech division. We will see innovation organizations face more noteworthy doubt and investigation from governments, representatives, and more extensive society. An expanding number of organizations will follow in the strides of Amazon, who as of late propelled its Sustainability Data Initiative, as they try to secure their ways of life as great corporate residents. Built up tech organizations will work to reinforce corporate informing with respect to the social and natural estimation of items and administrations, yet this methodology will probably crash and burn with educated partners except if it is supported by significant activity and expanded straightforwardness.

1. **Grasp Straightforwardness:** Absence of straightforwardness is probably the best hindrance to confide in right now confronting the tech segment. From sharing moral structures administering the advancement, what's more, arrangement of new innovations, to tolerating tax assessment in the nations in which they work, tech organizations need to grasp straightforwardness and a sense of metro obligation.
2. **Grasp Development:** Tech organizations have a task to carry out in building up the abilities of developments, yet accomplishing the capability of innovation can't be practiced by this segment alone. Organizations in different segments assume a critical job in interfacing mechanical aptitude with basic human needs. Just through this joint effort can the advantages of innovation be appropriated as broadly and as equally as could be expected under the circumstances.
3. **Comprehend Your Lifecycle Impacts:** The full lifecycle of numerous innovation applications and items is yet to be comprehended. Organizations must contribute to comprehend the impacts of advances–from advancement through to application. Just by doing so can these effects begin to be tended to.

3.2.5. Healthcare and Pharma

"Society can spend only so much money on healthcare, so we have to spend it wisely. It's important that we spend it on things that are actually creating value and having an impact."—Kenneth Frazier, CEO, Merck & Co.

From air contamination, to antimicrobial resistance (AMR), the soundness of people, animals, and the biosphere are inseparably connected.

The human services division faces concurrent difficulties and openings–from the dangers of AMR and atmosphere empowered pandemics, to the progressing customer-driven move to health and all-encompassing disease avoidance. Like never before, nourishment makers, retailers, pharmaceutical organizations and approach producers are being compelled to cooperate to discover answers for interconnected worldwide wellbeing challenges.

3.2.6. Lopsided Access

While financial and innovative advances are extending social insurance, it is spreading at a lopsided rate far and wide. Pharmaceutical and human services administrations organizations have the opportunity to help social insurance extension through joint efforts, development, promotion, and the mix of innovation into their existing plans of action to help United Health Care (UHC). Be that as it may, the industry likewise faces dangers of more tightly guidelines, tested development, and needs to discover its way in a world understanding the intensity of anticipation and progressively all-encompassing consideration.

China has stretched out essential wellbeing inclusion to in excess of 90% of the populace. The quantity of individuals ready to get key wellbeing administrations including vaccination, family arranging, and bed nets has expanded in the 21st century. However, over a portion of the total populace still can't acquire basic human services, regularly pushed into destitution because of paying for social insurance costs.

The biopharmaceutical business is participating in more multi-sectoral associations to help the development of UHC around the globe. For example, cell phones and video gushing administrations are as of now being utilized for remote specialist discussions furthermore, remedies.

The medicinal services segment is viewed as the area with the potential most noteworthy generally speaking effect on the achievement of the sustainable development goals (SDGs).

The ascent of constant diseases and a maturing populace will keep on moving human services organizations and governments towards more prominent interest in preventive medication, wellbeing, and an entire wellbeing approach (Figure 3.3).

Figure 3.3: Mobilizing capital will be one of the key success factors.

Source: WHO.

3.2.7. Sound Biosphere, Healthy Humans

From biodiversity to air contamination and environmental change to malady transmission and the development of antimicrobial opposition, our physical condition, imparted to animals, is inseparably connected to human wellbeing and the other way around. Biodiversity misfortune and environment harm—by the two people and a changing atmosphere—assumes a significant job in the spread of irresistible ailments and pandemics. As of late episodes of SARS, Ebola, flu, and intestinal sickness have been credited to human effects on biodiversity, untamed life exchange, or unsustainable land use.

The conservation of biodiversity is additionally essentially connected to sub-atomic decent variety and fruitful medication improvement. The mass annihilation of plant and animal species will have untold negative effects on human wellbeing.

The One Health idea perceives that the wellbeing of individuals is associated with the wellbeing of animals and the earth and has been progressively developing in conspicuousness among policymakers, industry, and different partners.

The financial estimation of products and ventures, including direct medical advantages, given by environments is evaluated at $33 trillion every year.

WHO evaluates that all-around in excess of 91% of the total populace are presented to air quality levels that surpass safe breaking points.

Because of expanding utilization of anti-microbials for people and animals and untrustworthy rehearses, anti-microbial resistance is turning into an undeniably earnest issue. Levels of medication safe contaminations are anticipated to cost the world $100 trillion in lost yield between now and 2050, which is more than the current worldwide economy.

Most of all anti-infection agents delivered are given to domesticated animals—about a portion of all anti-infection agents in the UK, 66% in the EU, and 70% in the US.

The US Food and Drug Administration has endorsed just nine anti-infection agents in the decade ago, and pharmaceutical organizations including Novartis and Sanofi have relentlessly stripped from their anti-toxins research to organize the advancement of progressively productive medications for incessant sicknesses.

3.2.8. Pandemic: Vigilance or Preparedness?

The worldwide network isn't doing what's needed to get ready for the following pandemic—with nations more powerless than any other time in recent memory because of changing populace socioeconomics, globalization, anti-toxin resistance, and environmental change. One hundred years on from the deadliest pandemic ever, Spanish influenza, gages have indicated that if a comparative occasion was to happen today, the loss of life could be as high as 147 million individuals around the world.

The global pharmaceutical market merits an expected $1 trillion; however, antibodies speak to just 3%. Vulnerability about when and where scourges will develop implies there is minimal money-related motivator for pharmaceutical organizations to put up genuinely necessary antibodies for sale to the public.

The world at present gives little to pandemic readiness, with Bill Gates cautioning it is the one zone where the world isn't gaining enough ground.

The US Center for Disease Control and Prevention has reported it will scale back its pestilence counteractive action exercises.

The UK as of late reported its first Vaccine Manufacturing Innovation Center to empower the fast advancement of prescriptions in case of the UK or worldwide plague.

Illness observation stays one of the most useful assets to help foresee, watch also, limit the damage brought about by flare-ups and expand information about what prompts pandemic episodes.

3.2.9. Future Expectations

Biopharmaceutical organizations will keep on concentrating on wellbeing advancement and aversion through advancements in item and administration advances which empower progressively customized, all-encompassing consideration. One Health will increase further noticeable quality, with various divisions, businesses, and partners imparting and cooperating to accomplish better general wellbeing results. Framework-wide ways to deal with handle anti-infection opposition will likewise develop in reach and significance in 2019, with more noteworthy joint effort between domesticated animals ranchers, nourishment makers, nourishment retailers, pharmaceutical organizations what's more, approach producers at neighborhood, territorial, national, and worldwide levels.

1. **Bolster Universal Health Coverage Endeavors:** Innovative work for future medications will stay critical for biopharmaceutical organizations and remain the biggest commitment they can make to UHC—through new medications as well as antibodies and other avoidance-related administrations. All organizations can ensure the advantages to their workers and all through their worth chains get us closer to UHC.
2. **Go About as One Health:** The Interagency Coordination Group on AMR met by the UN will share suggestions in 2019, which will control governments and industry on the bunch issues identified with AMR, including devices like guidelines. Ahead of time of that, life sciences organizations ought to think about joining the AMR Industry Alliance.
3. **Backer for Pandemic Readiness:** Governments are integral to planning for the following pandemic, yet business can likewise contribute arrangements including by supporting for more prominent impetuses required for interest in immunizations and associating patients and care suppliers with general wellbeing laborers through versatile media communications to identify and act against contamination episodes.

3.3. ENERGY AND UTILITIES

The energy segment has arrived at a point of change. Jolt is picking up force: from decentralized sustainable power source to electric vehicles (EVs), the upset has started.

The lines between oil and gas organizations and electric utilities are starting to obscure as conventional non-renewable energy source organizations put resources into the low-carbon change. Urban areas are on the forefront of progress—driving on savvy lattices, energy stockpiling, and EV foundation and digitized energy and carbon exchanges. Difficulties in regards to issues like foundation financing and fair access to reasonable clean energy stay focal.

3.3.1. Urban Energy Revolution

Urban areas are at the main edge of low-carbon developments and endeavors to turn away worldwide warming. Huge information and IoT keep on offering energy productivity gains for structures and transportation. Electric and self-sufficient vehicle innovation is prompting progresses in matrix stockpiling and decrease in emissions. Energy change is being demonstrated in urban areas, with exercises for locales and countries. Energy organizations will become accomplice at the city level to progress more brilliant, cleaner energy.

Worldwide spending on keen urban areas innovation is relied upon to practically twofold from $80 billion every year in 2018 to $158 billion out of 2021.

Chairmen have vowed that new structures in 19 urban communities, including London, Los Angeles, New York City, Tokyo, Paris, and Sydney, will have zero discharges by 2030, with existing structures getting zero by 2050.

Brilliant lattice advances could help cut ozone-depleting substance emissions by 3.9% by 2030.

Organizations including California utilities Pacific Gas and Electric (PG&E) and Southern California Edison are starting to address a "simply progress"–the evenhanded change to sustainable power sources and EVs.

EV possession is relied upon to arrive at 125 million by 2030, prodded by approaches that support drivers, armadas, and regions to buy zero discharges vehicles (Figure 3.4).

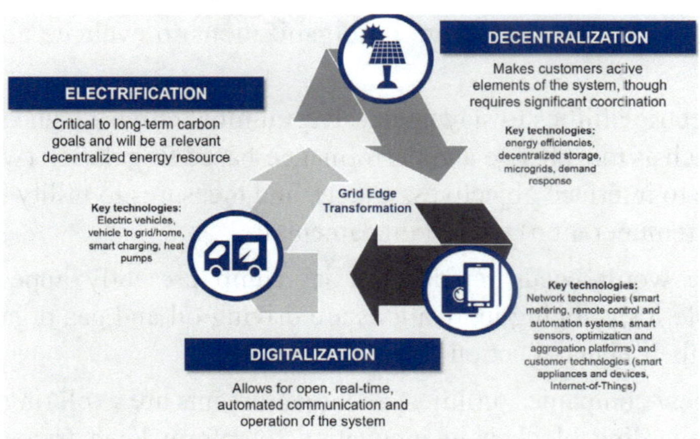

Figure 3.4: Electrification, decentralization, and digitalization.

Source: World Economic Forum.

3.3.2. Digitized and Decentralized Energy Trading

Jolt won't occur rapidly, yet its rising currently appears to be inescapable. Developing advances are constraining oil and gas organizations to change their plans of action to stay focused. Utilities will keep on moving ceaselessly from coal to renewables. These changes will require progressing and profound commitment with external partners, for example, financial specialists, government, and nearby networks.

Statoil (presently Equinor) and DONG energy (presently Orsted) have both experienced ongoing name changes that mirror their days of work away from petroleum derivatives and developing interest in sustainable power source.

Shell has been quickly putting resources into the power and EV space, with various acquisitions in the course of the most recent two years. The organization has focused on burning through \$1–2 billion every year until 2020 on low carbon arrangements.

French energy mammoth Total is planning to turn into an aggressive gas and electric supplier in Europe with its takeover of power retailer Direct Energy and overwhelming interest in sustainable power source and gaseous petrol.

Significant financial specialists, for example, BlackRock and State Street are starting to put open pressure on organizations to evaluate atmosphere dangers.

For electric utilities to stay aggressive, administrative advancement and change, such as rate change and Performance-based Regulation (which give a structure to interface objectives, targets, and measures to utility execution or official remuneration) will be fundamental.

Charge won't occur rapidly, yet its climb presently appears to be unavoidable. Developing innovations are driving oil and gas organizations to change their plans of action to stay aggressive.

Tech new companies, utilities, and governments are exploring different avenues regarding blockchain innovation to reform keen framework the executives and encourage shared energy exchanging.

Blockchain can empower little makers of energy to give Renewable Energy Credits (RECs) that would then be able to be followed and offered to organizations hoping to diminish their carbon impression. Blockchain empowered exchanging pilots are right now in progress in Thailand, Singapore, South Korea, Europe, the United States, Canada, and Australia.

The innovation will keep on being applied as our energy frameworks move to a more disseminated model. A developing number of battery organizations, including Californian battery new companies Sunverge and Stem, Panasonic, and German energy organization Sonnen bunch are utilizing blockchain to empower virtual distributed energy sharing systems that use home batteries and EVs to balance out lattice energy utilization.

Japan's TAKE Energy Corporation is offering clients a sun-based and capacity bundle which will enable distributed power deals through blockchain. The nation's second-biggest energy organization, Kepco, is sticking to this same pattern. SP Group, the power lattice administrator in Singapore, is permitting little scale sun-based and wind tasks to issue blockchain-produced RECs.

In the UK, a consortium of energy organizations is supporting the advancement of a blockchain-empowered adaptable exchanging stage for energy conveyance. Ben and Jerry's is utilizing blockchain innovation to break carbon credits—that balance GHG discharges of an organization's tasks—into small-scale exchanges that empower customers to purchase a carbon credit for each scoop of dessert bought.

3.4. FINANCIAL SERVICES

"CEOs will see in the tone of our questions that we are geared towards long-term performance—both from a shareholder perspective, but also from the perspectives of their stakeholder needs and their organizational resilience."—Matt Christensen, Global Head of Responsible Investment at AXA IM.

The money-related administrations area is at last paying attention to ESG also, not thinking back. Expanding consciousness of natural, social, and administration (ESG) issues, joined with the money part's developing comprehension of the effect of ESG on corporate execution, is driving banks, resource administrators and financial specialists to incorporate sustainability information into basic leadership at extraordinary levels.

ESG information has developed to where it has been grasped by the fund area as a key method to recognize the organizations situated to succeed, and those that may fail to meet expectations or fizzle. A larger part of ESG records outflanked their non-ESG partners in 2018. Feasible, mindful, and sway contributing resources under the executives additionally rose to $12 trillion in the US in 2018, up 38% from $8.7 trillion of every 2016. In 2019, expanding coordinated effort over the fund part will keep driving activity on themes like comprehensive free enterprise, environmental change, firearm viciousness, and assorted variety.

3.4.1. New Devices, Measurements, Alliances

From expanded interest in SDG-adjusted activities to new partnerships and alliances, monetary administrations organizations are awakening to the opportunity's introduced by a progressively practical economy. This year will probably observe more prominent investigation with respect to estimation of ESG and SDG-centered fund items. Elevated Level Expert Group on Sustainable Finance set up by the EU has distributed its proposals for a money-related framework that supports feasible ventures. In light of these proposals, the Commission is currently concluding a multi-year procedure on economical funds in the EU. The Embankment Project for Inclusive Capitalism (EPIC) has united CEOs from more than 30 organizations, resource administrators (like BlackRock, Fidelity, Vanguard, and State Street) and resource proprietors to recognize and make new measurements to quantify and show long term an incentive to money related markets. UNEP FI and 28 banks propelled the Principles for Responsible Banking to characterize the financial business' job and duties in molding a manageable future. A dominant part of ESG lists outflanked their non-ESG partners in 2018 as indicated by information from MSCI. Specialists anticipate that ESG records will keep on outflank a greater part of non-ESG during the worldwide financial downturn gage for 2019–2020. New items and instruments proliferate from Just Capital and Goldman Sach's ETF and Blackrock's ESG ETFs to the World Benchmarking Alliance SDG benchmarking; be that as it may, the effect of new venture items specifically is not yet clear (Figure 3.5).

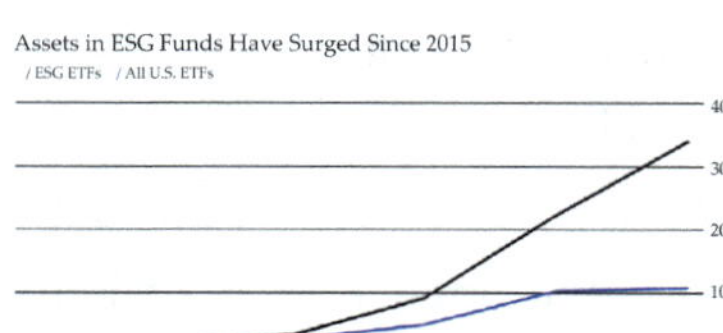

Figure 3.5: Climate risk and disclosure.

Source: Bloomberg.

3.4.2. Atmosphere Risk and Exposure

Driving banks and financial specialists are progressively incorporating atmosphere opportunity in their activities furthermore, are progressing from estimation to the executives. Banks and financial specialists are progressively working together to create measurements, draw in with organizations and request more prominent government activity on atmosphere.

UNEP finance initiative and 16 of the world's driving banks met up to pilot usage of proposals by the taskforce on climate-related exposures (TCFD) and create situations, models, and measurements to empower evaluation and revelation of atmosphere-related dangers and openings.

An ongoing report on organizations' readiness for TCFD proposals found that in the US, just 64% of money-related firms have board-level oversight of environmental change, contrasted with all organizations in the UK and 82% in Canada. As it were, 50% of organizations' environmental change opportunity administration systems expand further than six years out.

Around 415 financial specialists with a joined $32 trillion in resources have requested government's increment activity to handle environmental change, marking an explanation that asks governments to fortify their nationally determined contributions (NDCs) to meet the objectives of the Paris agreement.

The five-year Climate Action 100+ activity, presently supported by 310 financial specialists with more than $32bn in resources under administration, drives signatories to draw in with fundamentally significant ozone harming substance producers and organizations approaching them to improve administration on environmental change, control emissions and fortify climate-related budgetary exposures.

Portfolio decarbonization is getting a move on as increasingly institutional financial specialists seek after methodologies to dispense with stocks with high carbon risks.

3.4.3. Rising Shareholder Activism for Sustainability

Driven by more prominent attention to ESG dangers and openings, also rising desires from society and a more youthful age of financial specialists, resource chiefs, and banks are feeling the warmth and making more noteworthy moves on ESG issues. Prominent points like weapon brutality, assorted variety, environmental change, and narcotics are seeing the most activity among banks and speculation firms, driven by investor activism and progressively modern ESG risk examination. Resource directors are currently cooperating to support better rehearses among portfolio organizations on issues like sexual orientation assorted variety.

Goldman Sachs inquire about found that Twitter posts on ESG account themes have risen 19x since 2010. For the most part determined by exchanges identified with environmental change, venture-related terms (ESG, feasible contributing, sway contributing, and so forth.)

Florida Parkland secondary school shooting lobbyist David Hogg made a generally plugged call to blacklist BlackRock and Vanguard prior to 2018 because of each organization's interests in firearm makers and retailers. While the blacklist was ineffective, BlackRock, and state street both vowed to draw in firearm makers in discussions on expanding firearm security. BlackRock additionally propelled two new firearm free ETF files and bolstered lobbyist investor proposition requiring firearm security exposure, more–33x in 2017 from 2010 levels.

As you sow and CODEPINK discharged Weapon Free Funds, another device from to assist financial specialists with evaluating their record and shared reserve ventures for property in military temporary workers and weapon producers and retailers.

Further reacting to wild weapon savagery in the US in 2018, the CEO of speculation firm Calvert Research and Management, John Streuer, effectively constrained Kroger to quit selling guns.

Twenty-seven organizations (£10.5 trillion resources under administration) have now marked an announcement of purpose, composed by the 30% Club, to advance more ladies to the senior administration and sheets of the British organizations where they contribute.

BlackRock, state street, and vanguard are increasing weight on organizations to increment assorted variety and are starting to foundationally cast a ballot against male-just sheets.

Barclays set another open arrangement expressing it won't finance extends in World Heritage destinations or Ramsar Wetlands.

This year will bring activity, commitment, and results from solid alliances; for example, EPIC, the Principles for Responsible Banking, and the US Alliance for Sustainable Money—that framed in 2018. Budgetary administrations organizations that are not part of these collisions are probably going to be feeling the squeeze to join these organizations.

Banks will step-by-step start actualizing TCFD as they figure out how to survey and unveil atmosphere opportunity, while resource supervisors, financial specialists, and banks will keep on connecting with portfolio organizations on need subjects like atmosphere and sexual orientation.

We moreover expect a further deluge of ESG items and more prominent investigation with respect to the effects of those items.

1. **Work With Peers:** It is basic for money-related administrations to work and stand together to send an assembled sign to governments that ESG issues are basic to long term productivity also, budgetary achievement. Financial specialists and portfolio directors must begin inquiring vital ESG inquiries during organization profit calls and roadshows.
2. **Know Where You Stand:** With developing examination on corporate activities, banks, resource supervisors and speculators must be sure about their qualities and the moves they are making to genuinely epitomize their qualities.
3. **Put Resources into ESG Examination:** From human cash flow to AI and enormous information, having the aptitude and capacity to break down and assess ESG impacts, including atmosphere risk and opportunity, will turn into an upper hand for banks and resources chiefs.

CHAPTER 4

STRATEGIC MACROTRENDS AND DISRUPTIONS SHAPING 2020–2030

CONTENTS

In 2010, the World Business Council for Sustainable Development (WBCSD) delivered Vision 2050, a milestone piece of work that spread out a pathway to a world in which nine billion individuals can live well, inside planetary limits, by mid-century. In 2019, WBCSD chose to return to its Vision 2050 work–10 years on, there has not been as much advancement made along the Vision 2050 pathway as required. WBCSD is cooperating with 40 part organizations to refresh the pathway to mirror a portion of the incredible changes that have unfurled, to align the pathway with the Sustainable Development Goals (SDGs), and to focus on the basic moves that business can make to open the advancement needed for the sorts for changes that our Vision 2050 requests. From the second we chose to refresh Vision 2050, we realized we expected to consider what the following 10 years may toss at business. Not simply as far as sustainability challenges: we needed to research what the world would resemble, and consider what that would mean for business' capacity to work effectively (so they can likewise work economically). Our aggregate capacity to follow through on Vision 2050 (9+ billion individuals living great, inside planetary limits) is subject to our capacity to comprehend and explore a powerful working climate. This implies contemplating every one of the pressing factors that shape the world we live in, and that business works in, from socioeconomics and financial aspects, to innovation and legislative issues. WBCSD cooperated with Volans to research the world that lies in front of us.

The WBCSD research reports offer the proof behind our endeavor here to sum up the macrotrends and potential disturbances liable to shape that working climate throughout the decade ahead. This issue brief, and the exploration behind it, is anything but a bunch of gages, expectations or situations. We are making an effort not to put a specific twist–positive or negative–on the investigation. The fact is to take an unmistakable perspective on what the 2020s may have coming up, with the goal that we can foster more successful systems for how to react and gain ground on Vision 2050. We will require procedures that bridle the energy made by patterns that can go about as tailwinds, and we will require systems that are versatile despite unavoidable headwinds. One such headwind is the COVID-19 pandemic.

Very much like most different conjectures of ongoing years, we had distinguished a pandemic as something that the world would be unfit to manage were it to happen, also something that was quite liable to happen. But, we also wind up staggered by the appearance of the COVID-19 worldwide general wellbeing emergency in the primary long stretches of the decade. It is intensely exhibiting the significant, wide-running–and, usually, durable–

outcomes that such disturbances can have. Coronavirus has set off numerous troublesome stuns and sped up large numbers of the macrotrends that we had recognized. It has made certain things more probable and others more uncertain. With both the general wellbeing emergency and the connected financial disturbance actually unfurling, there stays a significant degree of vulnerability about what the full implications of COVID-19 will be for the decade ahead. In any case, the macrotrends and disturbances canvassed in this synopsis have not disappeared. As a rule, COVID-19 may serve simply to speed up or strengthen them. It is hence fundamental that we keep on thinking about the master plan simultaneously as arranging the reactions and inevitable recuperation from the current emergency. WBCSD provides a different research paper that glances in more prominent profundity at the ramifications of the COVID-19 pandemic, both as far as the macrotrends and disturbances spread out in this book and all the more comprehensively as to exercises gained from past recuperations and subsequently how business ought to react to this emergency.

4.1. MACROTRENDS

4.1.1. Demographics

The worldwide populace is developing, maturing, and urbanizing. By 2030, two-thirds of us will live in urban areas–and a developing number, especially in Africa and Asia, in megacities (however, megacities are relied upon to represent under 10% of the absolute metropolitan populace). Movement–both inside and between nations–is likewise expanding.

1. **Generational Handover: from Baby Boomers to Millennials and Gen Z:** Around the world, those brought into the world after 1980 now dwarf those conceived before. A generational handover of political, social, and financial force is as of now in progress and will proceed all through the 2020s. Progressively, pressures between the various ages are probably going to rise to the surface. These mirror the distinctive developmental encounters of various companions. For instance, the individuals who grew up around 2007–2008 will in general be significantly more doubtful of unregulated economy private enterprise than their archetypes who grew up during the Cold War. The nations with the biggest Millennial companions are all in Asia, the Middle East and Africa: India alone has 385 million millennials. Conversely, the nations

with the biggest baby boomer associates are in Europe, North America and Oceania: by 2030, generally 30% of the number of inhabitants in Europe will be more than 60 years old.

2. **Population Growth in Asia and Africa:** The populaces of numerous African and Asian nations are anticipated to keep rising quicker than the worldwide normal during the 2020s, adding to a continuous shift of international and monetary force toward these landmasses–and, maybe particularly, the developing number of megacities they contain. This will likewise put extra strain on regular assets, prominently water: four of the five most weak focal points for potential "water wars" are in Africa or Asia. The blend of rising flourishing and the rising recurrence and seriousness of environmental debacles is probably going to likewise make the quickly developing populaces of Asia and Africa altogether more versatile. Paces of relocation both inside and between nations are set to rise consistently. Thus, this is probably going to significantly affect legislative issues–both in nations that see undeniable degrees of displacement and in those that see significant degrees of migration.

4.1.2. Environment

As we enter the 2020s, human culture is as of now dealing with numerous planetary limits. Our monetary, social, political, and wellbeing frameworks all have an undeniable degree of interdependency with the climate. There is proof to propose that the blend of deforestation, biodiversity misfortune and environmental change add to an expanded danger of zoonotic pandemics. More slow movement natural emergencies, for example, soil corruption and pollinator misfortune, are sabotaging biological system administrations, diminishing the financial efficiency of key areas, like horticulture. As these different, interlinked emergencies extend, so the political and social craving for change develops. Public mindfulness and worry about sea plastics and environmental change flooded in the last part of the 2010s. Over the course of the decade ahead, we can hope to see more citizens and shoppers focusing on natural issues–especially where there is a solid, direct connection to their own wellbeing, similarly as with air quality.

1. **Worsening Climate Impacts:** As the planet keeps on warming, outrageous climate events will turn out to be more continuous and more extreme. This will cause enduring and relocation. It will

likewise force monetary expenses–harming property, subverting jobs and decreasing the efficiency of both land and work, with those in lower-income nations liable to endure lopsidedly. The probability of contention in districts that are antagonistically affected by environmental change will increment. Given this consistent drumbeat of awful news, tension on governments and organizations to convey significant environmental activity is sure to develop. Thusly, this will build the size of change dangers (and openings) confronting a few enterprises, as governments behind schedule receive more grounded environment arrangements. Furthermore, it's not just green legislators who will procure a political profit: so too will patriot strongmen who flourish with the sort of dread and precariousness that the environment emergency produces.

2. **Local Pollution, Degradation, and Scarcity Create Impetus for Innovation**: In spite of the fact that environmental change will without a doubt be the greatest natural story of the decade, other, more limited, ecological difficulties will likewise have critical ramifications for both business and society. Soil and land debasement will substantially affect horticulture in numerous districts. Air quality will be worrying in numerous urban communities–particularly the quickly developing megacities of Asia and Africa–with extreme ramifications for general wellbeing. 66% of the total populace could be living in water-scarcity nations by 2025. Like environment, these different biological difficulties will cause enduring, removal, precariousness, and financial expenses. In any case, they ought to, without a doubt should, likewise go about as an impetus for development–to accomplish more with less–and fuel requests on governments and organizations to oversee characteristic assets all the more capably.

4.1.3. Economics

COVID-19 has transformed the short-term economic outlook, but the longer-term trajectory of the global economy remains deeply uncertain. Given the weaknesses the global economy was carrying before COVID-19 hit (for instance, under-investment, high debt and inequality, slow productivity growth, stranded asset risk), the opportunities of a swift recovery are slim.

But the crisis also creates an opportunity for a political reset that could unlock a wave of sustainable and inclusive growth if policymakers rise to the challenge.

Coronavirus has changed the momentary economic viewpoint, yet the more drawn out term direction of the worldwide economy remains profoundly unsure. Given the shortcomings the worldwide economy was conveying before COVID-19 hit (for example, under-venture, high obligation and disparity, moderate efficiency development, abandoned resource risk), the odds of a quick recuperation are thin. In any case, the emergency additionally sets out a freedom for a political reset that could open a rush of manageable and comprehensive development if policymakers meet people's high expectations.

1. **Transient Crisis; Long-Term Slowdown**: The COVID-19 pandemic, joined with an oil value battle among Russia and Saudi Arabia has set off what appears to be the most noticeably awful economic crisis since the Great Depression. Estimates from March 2020 recommend that world GDP may fall by 4% or more this year–fundamentally more than the last worldwide downturn in 2009. Even before the pandemic hit, worldwide development was easing back and obligation levels were high. A few long-term factors, like low usefulness, high disparity and the danger of abandoned resources, made a log jam during the 2020s almost certain. Governments, organizations, and families are on the whole prone to rise up out of the pandemic with seriously debilitated accounting reports and expanded obligations, which could imply that there is a long monetary headache from COVID-19. The pandemic has additionally changed the political condition in numerous nations. Enormous monetary upgrade bundles are presently on the table in each country that can manage the cost of them. In the event that all-around focused, state intercession to launch monetary recuperation could trigger another flood of maintainable development as speculation is directed towards decarbonization and other long-term cultural objectives. In any case, it could likewise trigger a delayed time of something likened to 1970s-style stagflation. Also, there is a danger that pieces of the creating scene get abandoned, in light of the fact that they are essentially unfit to rescue themselves from the financial emergency.

2. **Pinnacle Globalization and the Rise of Asia**: The worldwide economy's focal point of gravity is moving conclusively towards Asia–and in a roundabout way to nations and districts that Asia (especially China) favors. A lot of worldwide GDP is figure to surpass half right on time during the 2020s and keep on ascending consistently. Right on time during the 2020s, Asia will likewise get home to half of the world's working class. With this comes a critical change in the locus of worldwide demand. Trade pressures will probably stay a component of the international scene all through the 2020s, prodding fracture, a shortening of supply chains and a reestablished center around public independence–may be sped up by the effect of COVID-19. Asset patriotism and rivalry between states for control of crucial crude materials are probably going to turn into an inexorably significant factor in global relations.

4.1.4. Technology

Coronavirus has sped up–and will keep on speeding up–the organization of a wide scope of existing advancements. Web-based business, distant working, internet learning and telemedicine have all gotten a lift that will just halfway disseminate after the emergency passes. The competition to foster an immunization is supercharging mechanical development in the drugs area. Robotization of creation and dissemination measures is speeding up as a method of limiting the spread of the infection temporarily–and to make supply binds stronger to the sort of stun a pandemic makes in the more extended term (however, amusingly, this may build weakness to various types of dangers, for example, cyberattacks). In the meantime, key advancements in various regions–from biotech to environmentally friendly power–are creating along remarkable pathways that could prompt financial tipping focuses during the 2020s. An essential inquiry is the way evenhandedly the additions of mechanical advancement will be shared. Numerous computerized advances made by developed nations brings home all the glory elements, which, without countervailing intercessions from government, will in general increment both market focus and disparity. There are a few signals that administrations may venture upon this issue during the 2020s: the EU's way to deal with antitrust requirement and the OECD/G20's work on corruption aversion are promising in such manner, however considerably more will be required if innovative advancement is to fuel comprehensive development as opposed to worsen disparities.

1. **Mechanization Impacts Every Industry and Country:** The spread of mechanization in each industry–from assembling to funding–will speed up, driven by the development of a wide scope of innovations like man-made brainpower, mechanical technology, and 3D printing. These can possibly empower more brilliant and more reasonable utilization of assets–and to fuel future development. Computerization will likewise cause disturbance for some specialists and put further strain on the connection among work and monetary security in numerous spots. One tradition of COVID-19 might be a reinforcing of social security nets, which would help alleviate the effect of robotization on specialists. However, assuming, then again, COVID-19 attendants in another period of starkness, there is a high danger that robotization will prompt rising disparity inside nations, filling outrage and disappointment, and fanning the flames of populism. Customary reskilling will turn into a prerequisite for most. Supply chains, as well, will be reshaped–and, as a rule, abbreviated–as admittance to leading-edge advancements turns out to be more basic for both assembling and administrations than admittance to a modest, talented labor force.
2. **Datafication: Smarter, More Efficient and More Surveilled:** For better or worse, business, money, and government will all turn out to be substantially more data-driven. This should empower enormous asset and work usefulness gains across enterprises—especially in regions like assembling, food creation, and coordinations——hat can possibly make riches and lessen natural effect. The money area will progressively saddle the influence of Big Data and AI to evaluate possible advances and speculations: if these are utilized to help survey sustainability impacts, this could help trigger a reorientation of economic streams on the side of practical turn of events. Also, we will definitely see new degrees of observation and control in assistance of the two benefits and force: 'Reconnaissance Capitalism' just as the Surveillance State. Without a cultural and administrative "techlash" that prompts substantially more tough implementation of antitrust and information security/proprietorship rules, Big Tech will keep on storing up residents' information and transform it into megaprofits. Without administrative changes, the advanced countries will retain all the glory market elements are probably

going to fuel disparity and smother contest. Algorithmic inclination could intensify the scenario by sustaining and demolishing disparities. Obviously, COVID-19 has sped up the organization of observation innovations in a few nations–from China to Israel. There is each opportunity that crisis estimates will become perpetual post-emergency.

4.1.5. Politics

There is not a single clear finish to be seen to the increment in polarization, populism, and patriotism that has been a characterizing normal for the second half of the 2010s. The underlying reaction to COVID-19 all throughout the planet has regularly featured the strength of patriotism and the shortcoming of multilateralism. There will be openings, in the outcome of the current emergency, to revamp a more powerful worldwide request, however at present, it shows up almost certain that administrations will turn their consideration internal, zeroing in on cultivating recuperation at a public level. Relations between states are probably going to turn out to be more value-based and, sometimes, more unfriendly. On a more sure note, the slow handover of political capacity to an age whose political perspectives have been fashioned in our current reality where the environment emergency and disparity are main concerns may prompt a lot more prominent prioritization of these issues by governments as the 2020s progress.

1. **Polarization and Radicalism on the Rise**: The social, environmental, and economic disruptions that lie ahead will lead to a world that is more polarized and a politics that is more radical. Levels of dissatisfaction with political and economic elites are high–and have been for some time: a 2017 survey of 18,000+ people across 23 countries found that 71% of people agreed with the statement "my government does not prioritize the interests of people like me." This will drive defenders of the status quo to embrace ever more reactionary positions. Meanwhile, those with less to lose will be increasingly open to radical alternatives. Just as the 1930s produced both the Third Reich and the New Deal, the 2020s will likely offer up a combination of resurgent populist nationalism and Green New Dealism.
2. **Geopolitical Instability:** The ascent of egalitarian patriotism all throughout the planet has essentially debilitated multilateralism as patriot pioneers will in general be antagonistic towards

worldwide foundations and deals. They likewise will in general overlook the intricacy of the advanced world and possess little energy for protecting the interests of partners or putting away cash and lives to look after security. Electorates in certain nations may rebuff egalitarian pioneers at the surveys for their absence of capability notwithstanding a general wellbeing crisis, however, somewhere else COVID-19 has plainly sped up a float towards patriot dictatorship. Also, COVID-19 is giving libertarian systems and organizations a pardon to twofold down on existing plans that were probably going to diminish worldwide strength (for example, the monetary and ecological liberation that we see occurring). While US-China contention rules features, there are numerous other likely wellsprings of insecurity: from Russia and North Korea to Iran and Venezuela. Countries in relative decay and those whose long-term possibilities are compromised by monetary, social, and natural change have less motivation to stick to the worldwide guidelines and standards that support dependability.

4.1.6. Culture

Social change will be generally molded by shifts in different regions. For instance, urbanization (and the expanded versatility of populaces inside and between urban areas) ought to speed up a change in social inclinations from possession to sharing. Coronavirus, in uncovering society's reliance, may make a few social orders become less individualistic and spotlight more on aggregate rights and obligations, especially according to wellbeing, cleanliness, and security. As the environment and biodiversity emergencies deteriorate, especially as far as their effect on lives and jobs, environmentalism, and natural equity's social and political clout is set to rise. In spite of the fact that in no way, shape or form widespread, this will change perspectives to what and how we burn through for a huge minority of the worldwide populace and may trigger at any rate a halfway shutting of the hole between what buyers say about sustainability and how they act. Rising degrees of movement (coming about because of environmental change or different variables) may likewise fuel political disunity, since migration is frequently a central point of interest for libertarian patriots, regardless of whether against foreigner legislative issues regularly has little to do with real migration numbers.

1. **Materialism and Post-Materialism:** Attitudes and Lifestyles Diverge: The manners by which people measure their status and achievement fluctuate significantly across nations. 70% of Chinese say they measure their prosperity by the things they own. Just 21% of Swedes and Spaniards agree. In created economies, and especially among more youthful age partners, more individuals are beginning to esteem encounters and access over responsibility for products. Ways of life are adjusting to a developing cognizance of planetary limits–at any rate among some segment gatherings. Interest in wellbeing, good dieting, personal satisfaction and personal development is on the ascent, and a portion of the working class practices of mass isolates all throughout the planet will acquire ubiquity. The consumerist way of life holds a solid allure in numerous places where individuals have not yet gotten an opportunity to appreciate it, in any case. Financial aspects isn't the sole determinant of this disparity: in Peru, only 30% of individuals measure their prosperity by what they own, which is generally identical to the levels in Germany (31%) and Australia (29%)–nations whose per capita GDP is just about multiple times higher than Peru's.
2. **Culture Wars Escalate**: Qualities divisions are developing and hardening inside numerous countries all throughout the planet. Polarizing occasions–from the Brexit vote in the UK to new Citizenship Laws in India–are both a manifestation and reason for these partitions. A spellbound and divided (web-based) media scene additionally intensifies the pattern. There is no single issue that characterizes the new culture wars: they are about mentalities to race, religion, movement, sex, sexuality, place, globalism, nationhood, and that's just the beginning. These social parts map defectively onto segment partitions: in numerous nations, there is a metropolitan rustic as well as old-youthful measurement to these culture wars; frequently schooling level is a solid indicator of individuals' social perspectives. The way of life wars add to political polarization and thusly feed off it. Given this supporting input circle, there is a danger that conflicts over personality will increase in the decade ahead.

4.1.7. Significant Level Ramifications of the Macrotrends

How might these macro trends affect Vision 2050? Unquestionably, some can possibly brush us off base if not very much oversaw. Yet, others could practically be outfit to speed up progress. Here we research six of the patterns (grouped into three themes) that are general as far as their ramifications and which bring up significant vital issues for the Vision 2050 plan.

4.1.7.1. The Politics of Generational Handover

The rise and fall of distribution ages is a steady in human culture, so what makes the generational handover that will unfurl during the 2020s prominent? Old and youthful consistently see the world in an unexpected way, yet, in some significant ways, the interests and assessments of the present old and youthful are further separated–and possibly in more prominent pressure with each other–than in earlier many years. For millennials and Gen Z, their early stages have been overwhelmed by the aftermath of the 2007–2008 crisis–and now the COVID-19 pandemic. They have grown up with the mindfulness that environment mayhem in the course of their lives is almost certain except if revolutionary changes to the worldwide economy are executed. This setting has brought about numerous millennials and Gen Z being profoundly embittered with the state of affairs (counting with 'private enterprise') –however, there are special cases for this, outstandingly in quickly developing economies like Indonesia. While all ages will in general get more moderate with age, there is a decent possibility this pattern will be less articulated among those at present under 40, exactly on the grounds that they have been financially pressed and realize that, on current directions, they will experience their advanced age in a world seriously disturbed by environmental change and biodiversity misfortune. The political results of this specific generational handover could along these lines be significant–however, it isn't sure how this will work out during the 2020s. There are three conceivable situations:

1. Recent college grads and Gen Z neglect to enlist as a huge political power during the decade–in light of the fact that they "conform," don't turn out at decisions dependably enough and additionally in light of the fact that they are excessively divided. In this situation, the 2020s are overwhelmed by the undeniably moderate governmental issues of the children of post-war America. Another variety of this situation includes millennials specifically being "skipped," as boomers keep on sticking to

control from one side of the planet to the other. This could prompt a quick ascent in the impact and force of Gen Z, as they assemble to try not to be abandoned similarly as the recent college grads before them.

2. Twenty to thirty-year-olds and Gen Z are a destabilizing power because of significant degrees of estrangement with business as usual. In this situation, their impact makes governmental issues more limited and energized; bargain, alliance building and multilateralism all become harder.
3. Twenty to thirty-year-olds and Gen Z become the appointive bedrock of viable reformist governments, which focus on issues like financial change and tending to environmental change. In this situation, they channel their disappointment into promotion for administrative intercessions that location social imbalance and natural harm. From the viewpoint of Vision 2050, situation 3 is obviously the most ideal. Vitally, however, it requires the two organizations and governments to exhibit that a valid reformist elective exists.

4.1.7.2. Automation, Jobs, and the Economy

In numerous nations, the connection between work and financial success has been debilitating for an age, with a thin monetary first-class catching an outsized portion of the new abundance made by usefulness gains. A mix of powers–mechanical change, globalization, the declining force of worker's organizations and a persevering spotlight on investor esteem among numerous organizations–has prompted stale wages and progressively unstable types of work for some specialists. From gig economy laborers in the US to traveler workers in India, the financial effect of COVID-19 is hitting a considerable lot of these equivalent individuals hardest. Innovation and exchange may make social orders more well off in total, yet, without viable rearrangement, they aggravate a few people off–either in outright terms or comparative with different districts and financial classes. This has been an essential factor in the ascent of populism, patriotism, and the dismissal of globalization–prominently in the US, yet in addition somewhere else. Lamentably, the monetary arrangements being sought after by egalitarian pioneers all throughout the planet today are probably not going to change this–and, in certain occasions, will compound the situation. Exchange wars will dispense monetary damage on the individuals who have profited by globalization (in both created and non-industrial nations), without essentially improving life for the individuals who have not. Elective ways are conceivable. Computerization will keep on making interruption occupations, yet organizations and governments could act to alleviate the unfavorable cultural effects of this pattern by building more grounded social wellbeing nets, reallocating riches, and moving to a more partner situated type of free enterprise. Coronavirus has set off a brief reinforcing of social security nets in certain nations. Governments have carried on of need and a considerable lot of the crisis estimates embraced in the early long stretches of 2020 are not monetarily practical over the long haul. Yet, by making the already inconceivable inescapable, COVID-19 may have improved the probability that laborers, bosses, and governments will discover better approaches to share the dangers and prizes related with mechanization all the more fairly.

4.1.7.3. Nationalism and Globalism

The resurgence of political and social patriotism that has happened throughout the most recent decade gives no indication of blurring. This has debilitated multilateralism–and will very likely keep on doing as such. Having libertarian patriots in power builds the danger of war breaking out among countries, and it makes a setting where rough fanatics inside society feel encouraged. A customary way of thinking would recommend that this is terrible information for the Vision 2050 plan–and without a doubt from numerous points of view it is. Sustainability issues (especially environmental change) have to a limited degree speed in the way of life wars that are seething in numerous nations. Supporters of sustainable improvement are frequently seen (without anyone else just as by others) as globalists–individuals who esteem their way of life as "residents of the world" profoundly; who are prouder to be "woke" than to be energetic. There are, be that as it may, openings just as dangers for the Vision 2050 plan in this, especially if the accomplishment of reasonable improvement needs can be hitched to sensations of public personality and pride among residents, and to political goals of public safety and intensity. Parts of the Vision 2050 plan will in any case require multilateralism and business should attempt to reinforce multilateral cycles and foundations where it can. Yet, a significant part of the groundbreaking change required can occur at a public or even sub-public level, and keeping in mind that this may make intricacy for worldwide organizations that work across various wards, it might likewise set out open doors for quicker advancement in numerous spaces.

4.2. DISRUPTIONS

Not everything on this rundown will happen during the 2020s, yet all are conceivable enough that they warrant pondering and, to the degree conceivable, planning for. Not all are terrible–and, even those that are can possibly catalyze positive changes, similarly as past emergencies have. As with the macro trends, there are numerous potential interlinkages between these interruptions. Cascading types of influence–where one disturbance triggers others–are one opportunity. For instance, we are as of now living through a worldwide pandemic that may yet trigger a monetary emergency, mass unemployment, a significant clash as well as famous uprisings that lead to a shift in power. In the marginally longer-term, COVID-19 may likewise have improved the probability of a tipping point for the worldwide energy change showing up during the 2020s (surely a few examiners currently accept that petroleum derivative interest may have crested in 2019), which could thus trigger an unexpected rectification in the manner finance business sectors value environment risk. The way that the tech monsters will more likely than not rise up out of the current financial emergency in a much more grounded position comparative with (potential) contenders may build the credibility of an administrative kickback later in the decade.

In the interim, expanded familiarity with the connections between agribusiness, deforestation, biodiversity misfortune and the danger of zoonotic pandemics like COVID-19 could give an after wind to the biotech business, speeding up the change in food creation from land to lab. At long last, while COVID-19 may in the present moment have occupied worldwide pioneers from managing environmental change, it might likewise have created an enduring change in residents' assumptions regarding the job of government in reacting to emergencies and interceding in monetary life to guarantee results that advantage society. Joined with the requirement for significant public venture to invigorate monetary recuperation, this could make the conditions for something like a Global Green (New) Deal to arise during the 2020s. To put it plainly, while none of this is ensured, the COVID-19 pandemic could yet go about as a trigger for each and every other interruption on this rundown during the decade ahead. Kindly note, we have given a different Vision 2050 research paper investigating the drawn out results of COVID-19 in more profundity.

4.2.1. Financial Crisis

The joined impact of nations going into lockdown to moderate the spread of COVID-19 and an oil value battle among Russia and Saudi Arabia has hastened one of the quickest, most profound financial exchange crashes ever–and that is simply Q1 2020. Dissimilar to in 2007–8, when money related arrangement (e.g., quantitative facilitating and financing cost cuts) for the most part took the strain of restoring the worldwide economy, governments have effectively needed to carry out uncommon monetary measures in light of the Covid prompted crash–and there will be substantially more boost to come. A great deal is riding on how very much focused on government bailouts and boost bundles are–including how versatile the economy is the point at which the following accident comes, as it definitely will.

4.2.2. Global Pandemic

At the hour of composing, the quantity of affirmed instances of COVID-19 keeps on rising dramatically in numerous nations all throughout the planet. It is probably going to keep making death toll and interruption social orders and economies for quite a long time–if not years–to come. Furthermore, COVID-19 didn't appear unexpectedly. The previous 20 years have seen five possible pandemics–SARS, H1N1, Zika, Ebola, and MERS–which were all just a single hereditary change away from being pretty much as troublesome as COVID-19. We may yet see further pandemics in the decade ahead, with similar potential to overpower medical services frameworks, disturb supply anchors and shut economies and social orders for a while. Obviously there are exercises to be discovered that could improve versatility notwithstanding the following pandemic. Regardless of whether governments and organizations notice those exercises stays not yet clear.

4.2.3. Major Conflict

The opportunity of a genuine struggle between at least two countries with incredible militaries and abundant resources can't be precluded. Were such a contention to break out, all things considered, digital assaults on basic framework would highlight vigorously in the strategies of the fighting countries (in fact, ostensibly, there are now numerous cyberwars seething all throughout the planet). During a time of developing digital reliance, this could cause extremely critical damage and death toll among customary residents. So too could the organization of atomic, natural or compound weapons. As occurred in the Second World War, such a contention would probably trigger a full preparation of society and the economy in help of public conflict points in those nations included–and would create a feeling of social fortitude inside and between countries unheard of since the 1940s.

4.2.4. An "Economic Singularity" Leading to Mass Unemployment

Assessments about the effect of computerization and man-made brainpower on the quantity of occupations in the worldwide economy shift so uncontrollably that it's ideal to take any figure with a huge touch of salt. However, imagine a scenario in which those determining an inevitable breakdown in the absolute number of occupations—here and there alluded to as a financial peculiarity—are correct. Regardless of whether, over the long haul, innovation creates many new positions as it annihilates, the danger of certain locales and areas enduring a delayed time of extreme unemployment is high. Also, as the COVID-19 emergency illustrates, mechanization isn't the solitary factor that can trigger a time of mass unemployment. The danger of a delayed time of mass unemployment—regardless of whether because of computerization, a pandemic or another factor—has huge ramifications for political strength: governments will probably battle to adapt to the aftermath and this could fan the flares of revolutionary legislative issues and populism of both left and right.

4.2.5. "Techlash"

In the 20 years since the dot-com bubble burst, the tech area has had a simple ride. Social orders have accepted new advancements with insignificant dithering, and controllers have tossed moderately couple of barricades in the method of tech organizations doing what they need. That may now be evolving. There is rising disappointment and disappointment in numerous nations with the manner in which computerized innovations are being utilized to serve the interests of a limited monetary as well as political tip top. Customers are more mindful of the adverse consequences of innovation on wellbeing and prosperity. Worries about protection and information possession are on the ascent, however governments are not generally as worried about security as their residents are. Contest requirement has become an undeniably hotly debated issue in the EU and the US. New principles about the lawful status of "gig economy" laborers and the duties of web-based media organizations for the substance they host could subvert the reasonability of stage plans of action. And afterward, there's the effect of international contention: the furore over Huawei could be a preview of what might be on the horizon, with governments boycotting unfamiliar firms and unions stressed because of picking sides.

4.2.6. Popular Revolts Leading to Regime Change

Rising disparity inside nations and an apparent absence of responsiveness from political elites is powering an assorted arrangement of (generally peaceful) fight developments all throughout the planet. Given that the hidden reasons for these well-known uprisings look set to deteriorate during the 2020s, almost certainly, both the recurrence and seriousness of fights against business as usual will rise. The crucial obscure is the way viable they will be: the 2010s saw a sensational decrease in the achievement paces of peaceful missions requesting foundational political change. But there is no assurance that this pattern will proceed: as residents become angrier and more estranged, we could see apparently stable political systems across the world either brought down or compelled to act considerably more unequivocally in light of residents' requests.

4.2.7. A Climate "Minsky Moment"

Throughout the 2020s, there is little uncertainty that finance business sectors will improve at estimating both progress and actual environment risks into resource valuations. Other natural, social, and administration issues are additionally progressively on the finance business' radar, yet the environment is the issue destined to prompt a quick and emotional repricing of resources soon. There is a decent possibility that environment-related monetary revelations will get compulsory at some stage in the main portion of the decade. Innovations that empower more prominent straightforwardness into resource level information are likewise growing quick. As the nature of data accessible improves and the expected misfortunes to financial backers become both bigger and more inescapable, a significant–and possibly abrupt–reorientation of economic streams turns out to be progressively likely. This could trigger a purported environment "Minsky Moment": an unexpected descending amendment of resource valuations in carbon-concentrated areas.

4.2.8. Energy Transition Reaches a Tipping Point

Energy progress watchers are isolated into two camps: those that expect continuous change and those that anticipate a significantly more fast innovation as well as policy-driven disturbance over the coming decade(s).

If the last gathering is correct, at that point, we can hope to see interest for petroleum derivatives pinnacle and start to fall during the 2020s, which would have significant ramifications for the energy business, financial backers and, obviously, environmental change. The speed and size of cost decreases in the sustainable power area in the course of the most recent decade have reliably surpassed assumptions, and renewables are presently at or near cost intensity with non-renewable energy sources in numerous pieces of the world. Depending on how rapidly those expense decrease bends base out, a financial tipping point might be nearer than most experts might suspect. This doesn't imply that the switch will occur incidentally: it will in any case set aside some effort to work out limit and foundation, and governments and organizations that keep on putting resources into petroleum derivative limit today (for instance, China's financing of coal power creation) will attempt to try not to record abandoned resources as long as they can.

4.2.9. Biotech Boom

Engineered science can possibly change food creation, medication, and materials production. In a portion of these spaces, we may see nonlinear improvement during the 2020s as costs decay dramatically. For instance, one ongoing report on the food business anticipated that meat created in a lab, utilizing a cycle called 'exactness maturation,' will get cost cutthroat with conventional hamburger from cows in the mid-2020s and could be really much significant for the farming business, just as for human wellbeing and the climate. Huge parcels of agrarian land could be opened up for different employments. In the meantime, in the wellbeing area, engineered science is being utilized to foster quality treatments and customized prescriptions, while the plastics business is progressively putting resources into bio-based other options.

4.2.10. Global Green (NEW) Deal

Actually, like the 2000s, the 2010s got done with a disillusioning arrangement of environment talks (COP15 in Copenhagen in 2009; COP25 in Madrid in 2019). In any case, there is a critical possibility that the 2020s will see a leap forward in worldwide environment strategy activity–however, whether this will occur through the systems of the Paris Agreement or not remaining parts an open inquiry. From the EU's Green Deal to South Korea's declaration of a net-zero objective, energy is working in huge economies across the world. This is notwithstanding negative political changes in some key nations (e.g.,

US, Brazil, Australia) since the Paris Agreement, and the proceeded with opposition of states like Russia and Saudi Arabia. While the last are probably going to stay impervious to environment strategy activity for a long time to come, there is each opportunity that, post-trump, the US will venture (once more) into a position of authority on worldwide environment activity during the 2020s. Across the globe, resident activism is probably going to be a critical driver of political activity—and the consistent drumbeat of outrageous climate occasions will keep environmental change high in open cognizance and on the political plan for most governments.

CHAPTER 5

CREATING RESILIENT STRATEGIES

CONTENTS

5.1. STRATEGY CREATION

At the point when organization CEOs are pulled before administrators or researched by the press to clarify their duty issues, work rehearses or ecological calamities, something has plainly turned out badly. Additionally defective are the organizations that pass up new business sectors since they neglect to see the potential in web-based media, natural insurance or developing business sectors' thriving working class. The chiefs of such organizations have neglected to foster procedures that are adequately far-located and vigorous to adapt to a quick changing working setting. Albeit barely any hit the features as fantastically as Starbucks, Apple or BP, many organizations have been similarly blinkered about urgent changes on the planet that influence their systems.

We accept that essential reasoning is commonly too thin, caught in ordinary investigation and incognizant in regards to troublesome changes in individuals' needs and standards. Firms work in an intricate, unique setting–including monetary, social, political, innovative, social, and ecological cooperations. They need to completely comprehend the idea of this unique situation, tolerating that organizations need to deal with a progression between associations with other key entertainers in the public arena. They are a necessary piece of this greater picture. Fostering this more extensive viewpoint, extending skylines and time periods to spot openings for development that will support long-term achievement is what we have called Resilient Growth.

A few organizations are aware of quick monetary movements and conceivably troublesome innovations in their businesses, yet few appear to be prepared to search for long-term problematic patterns. As of late, the innovation area has created instances of development with a capable outlook and versatility. Organizations, for example, Patagonia base their constitution on various qualities and aspirations. Be that as it may, most settled organizations are battling to re-set the conventional plan of action and vital setting. Accordingly, essential deduction inside many organizations botches fundamental freedoms for development.

The risks of botched freedoms and powerful dangers are more prominent now than at any other time. These are unpredictable occasions. Organizations need to foster techniques with regards to extraordinary vulnerability about what's to come. Organizations must be adequately prepared to respond to advancements, while as yet seeking after a system that guides dynamic towards long-term objectives inside a structure of crucial qualities. Vital reasoning is planned to deliver an incorporated viewpoint of the business, however on the off chance that it overlooks the amazing powers influencing an organization and its business sectors, it can't satisfactorily distinguish the heading where the association should move. We accept strong procedures are expected to give the more profound bits of knowledge that will bring long-term achievement.

5.1.1. How Ready Are Organizations to Key Nearsightedness?

Accepting that there is a requirement for a major change in the manner organizations advance to guarantee a long haul, reasonable future, we explored how and why a few organizations have started to foster strong techniques.

1. Is an organization's business procedure advancement educated by the monetary, social, political, innovative, and ecological setting where it works?
2. How does the organization endeavor to recognize problematic turns of events and keep away from the snare of straight reasoning?
3. How does the business endeavor to comprehend its interrelationships with numerous partners and handle clashing requests?
4. What makes for progress or disappointment in essential reasoning? We trust this exploration will incite banter about the significance and importance of the more extensive cultural setting in creating effective techniques?

5.1.2. Instability and Vulnerability

The beginning stage for our request was to address how organizations can predict possible dangers and opportunities that should impact their essential bearing and arranging measures. Business experts realize they can't anticipate the future yet they should endeavor to get ready for what lies ahead. This is more troublesome as the world is changing in a heap of ways at a frantic speed, for certain principal changes lying outside the ordinary extent of corporate aptitude. Corporate Citizenship has outlined four megatrends that organizations need to get ready for (Figure 5.2).

Societal threats and opportunities

The business context is Volatile, Uncertain, Complex and Ambiguous (VUCA). This applies to the social, cultural and political components as well as to economic and technological aspects. We see four megatrends that companies need to be aware of in developing strategy:

Crunch – pressure on the availability and affordability of resources ranging from food to precious metals, putting a premium on sound stewardship

Fragment – splintering of traditional governance and social structures, growth of new organisations, expectations, attitudes and activities, requiring multi-dimensional relationships

Connect – new ways of sharing and storing information, bringing people together, increasing scrutiny and threatening privacy, demanding transparency and responsibility

Re-balance – the growth of new economies and new markets with different characteristics and priorities, requiring new approaches

Companies with resilient strategies that respond to these trends will find new opportunities, such as closed-loop products, fruitful relationships with customers and communities, and services for excluded groups. Those that do not will find themselves losing competitiveness, suffering censure and missing out on growth markets.

Figure 5.1: Social threats and opportunities.

Understanding advancements, for example, moving force structures, new friendly connections and incredible new needs is imperative for organizations to foster fruitful methodologies. However, our exploration proposes that even organizations among the CSR pioneers are attempting to coordinate these elements in their endeavors to adapt to the dubious future.

5.1.3. How to Adapt to Vulnerability?

There are no straightforward responses to the difficulties of vulnerability, instability, and vagueness about the future business, social, political, and social climate. Notwithstanding the troubles in any case, organizations that fall into the snare of extrapolating from ongoing experience hazard experiencing huge shocks whenever startling occasions or openings arise. Indeed, even all that laid plans can't expect the 'obscure questions,' however tacticians need to consider a wide scope of improvements and follow up on those that are the most critical and material to the organization. Our exploration discovered proof of such a wide-based skyline examining among a few of the organizations we met. In 0the most complete model, one organization has made an exploring bunch that counsels a wide scope of assessment formers and specialists in a few disciplines not straightforwardly identified with the business. The point is to accumulate insight on the key social and logical movements that will shape the eventual fate of the business. Another organization has a Strategic Foresight Team which inspects the worldwide patterns that may affect the business over the more drawn out term, educated by a course of internal commitment with chiefs.

Chief administration is tested to reevaluate their major perspectives on how the world and the cutthroat climate are developing. The methodology interaction centers around individual themes and thinks about situations for conceivably problematic powers. Situation practices are utilized in different organizations, with an essential suspicion that there will be huge changes. In one case, the senior authority group programs conversation of long-term advancements at its yearly gathering. While such activities help to recognize dangers and openings, one interviewee remarked that, at last, there would consistently be vulnerabilities, and it is important to accept that the central methodology of the business will support achievement. We additionally discovered proof of less deft practices. In a few organizations the methodology cycle has all the earmarks of being fairly unthinking and tied near more quick business drivers. Interviewees enlightened us regarding brand insurance work, statistical surveying and quick issues in their business sectors, however not with regards to more extensive and less prompt angles. Some expressed that they have a drawn-out point of view; however they couldn't well-spoken what this has really meant for the business system. Others remarked that the CSR groups know about significant long-term patterns, yet these don't show up in the danger register and openings network that advises key reasoning. In one case, the CEO has encouraged his groups to search for troublesome impacts; however there is all things considered a solid accentuation on investor esteem in arranging and a period skyline of only three years.

5.1.4. How Wide Is the Viewpoint?

Organizations that are modern in their way of dealing with vulnerability and non-direct reasoning may by the by have a genuinely thin spotlight on regular business drivers. For instance, innovation organizations realize that interruption is common in their area, however, even their all around created innovation filtering may in any case miss key cultural patterns. A solid beginning stage for a more extensive way to deal with methodology is vision and qualities which perceive the significance of cultural connections. We noticed this methodology in a few organizations we addressed. In two organizations the qualities were set somewhere near the author/proprietors and are unmistakable in all major decision-making. One current CEO demands that all ventures should stick to the establishing esteems while one more discussions about "another mechanical upset" and the "associated world" the organization is important for. Another tells representatives that in spite of monetary pressing factors, think about the organization "filling a more significant need," which is reflected in the upsides of reasonableness, uprightness, and straightforwardness. The vision is a beginning stage, yet it should animate reasoning that accepts society past the limits of monetary measurements and the momentary requests of the market. At the end of the day, it should be a live, significant thought instead of only an assortment of fine trademarks. It should drive chiefs to draw in with the more extensive setting of financial, social, political, mechanical social and ecological connections. The vast majority of these organizations endeavor to comprehend cultural patterns through their customary associations, in some cases including formal partner commitment. Clients, particularly those with high-profile buyer openness, are a significant wellspring of qualities arranged contributions for one organization.

However, while a few organizations effectively examine for such turns of events, some seem to depend on routine statistical surveying, which is regularly limited to standard business issues and buyer perspectives, conceivably missing significant turns of events. One interviewee portrayed searching explicitly for joins between friendly conduct and innovation and the organization's items. Be that as it may, one more said the organization depends on the corporate danger work, which was accepted to have excessively restricted of a viewpoint.

5.1.5. Broadband Reasoning

Attempting to expect what's to come is the beginning stage. The everyday the truth is that this interaction needs to squeeze into an authoritative construction. The organizations remembered for our examination revealed to us how to make the suitable administration detailing lines that help the advancement of long-term thinking on business system. We were especially intrigued by how they coordinate cultural viewpoints into methodology, with or without the investment of CSR experts and specialists.

5.1.6. Who Is Mindful?

Extreme obligation regarding technique in each business lies with the CEO and leader group. In a large portion of the organizations we addressed, procedure advancement is designated to others outside this chief group, however in one case ranking directors structure a Strategy Council that meets monthly to audit the vital parts of the business climate. Most organizations in our example have devoted system or business improvement groups who lead on technique advancement, however we discovered improvements both away from and towards this model. One organization has as of late delegated a Director of Strategy to foster a more noteworthy vital core interest.

For another situation, the committed corporate procedure work has been disbanded as it was viewed as having an "ivory tower" approach. All things being equal, each significant capacity in this organization currently has key liability regarding its own subject matter. In different organizations, every business division is answerable for its methodology, with the middle having a coordinating and evaluating capacity.

5.1.7. What Sort of Hierarchical Core Interest?

Organizations can embrace various ways to deal with system improvement, creating methodology for the whole organization or procedure for a specialty unit. Most organizations we met revealed to us that they foster technique in both the corporate and specialty unit setting. In one case, be that as it may, a group has been made to zero in explicitly on a significant new essential drive at a more miniature level. The unit is treated as a smaller than normal beginning up inside the business, answering to the technique group. This Insight Team is entrusted with investigating new market openings, understanding mechanical turns of events and making imaginative items and administrations that will fulfill expected shopper needs in 10–15 years' time. The thought is to make a solid spotlight on explicit goals, liberated from large numbers of the standard corporate limitations, to accomplish outstandingly quick advancement. This Insight Team is tied in with making a move to foster new ideas (in this specific case planning a drastically new item), not just distinguishing future difficulties or market requests. Note that this methodology is distinctive to the Strategic Foresight Team at another organization which focuses thinking on future patterns.

5.1.8. What Is the Time Skyline?

We tracked down a wide scope of ways to deal with the broadness and profundity of vital reasoning. A few have long timescales and expansive skylines, however others seem to consider procedure as minimal more than medium-term arranging. Five-year timescales are average. Three organizations are from the utilities sector so their methodology is normally determined by the administrative structure and timescale. Generally, this envelops a five-year plan as well as a prerequisite for a 25-year key vision.

In the innovation area, skylines can be long-term (consider Google Glass and Amazon drones), but simultaneously can be likewise compacted by the speed of mechanical change and the generally short lifecycles of items and administrations. Two organizations have a 20-year viewpoint and two others expressly request that chiefs think ten years ahead. In one model, a ten-year procedure was presented a couple of years prior on the grounds that the CEO saw the need to consider factors past the quick, transient issues that engross ranking directors. Outside of these models, nonetheless, there were nearly as many organizations where it is uncommon to look forward over five years.

5.1.9. What Sort of Information Sources?

Outside inputs assist chiefs with acquiring a more extensive viewpoint. Technique counseling firms were regularly referenced as suppliers of external ability and knowledge. Industry bodies additionally give area explicit data sources and a few organizations utilize free statistical surveying and financial examination. Just one organization approaches an outside counseling organization to get CSR skills and a more extensive cultural setting to the technique advancement measure. One method of social event various perspectives is through a conventional board or warning board. One organization has such a gathering comprising of senior leaders from different ventures, the scholarly community, non-governmental associations (NGOs) and different specialists. It meets double a year with ranking directors. Another has made a leading body of specialists on water assets–the key regular asset pertinent to the organization's area. The specialists assist ranking directors with understanding the basic external improvements to which they need to react. Another organization teams up with a scholarly establishment that spotlights on wide issues for the fate of the area.

5.1.10. Is There An Association Between CSR and Procedure Groups?

While the corporate vision oftentimes consolidates cultural aspirations–expressly in a few cases–this isn't really reflected in a nearby association between the CSR group and the procedure interaction. In something like three cases the head of CSR plays a significant part in procedure advancement. Nonetheless, in many organizations, the job of the CSR chief is more roundabout. Their perspectives and mastery are frequently separated through one of the senior chiefs, who goes about as a CSR champion, or senior leaders might consolidate cultural thoroughly considering their enrollment in a CSR Council or Committee. One interviewee disclosed to us that CSR and vital arranging had been completed autonomously previously yet there were presently moves to adjust the two. Any oversight of this more extensive logical deduction in the system advancement measure is twofold edged. It implies that CSR experts are not adequately drawn in with corporate methodology, just as the business may possibly neglect to assess key cultural advancements for their potential benefit.

5.1.11. Business and Society Connections

The third region our request considered was the manner by which the methodology improvement measure is educated by external viewpoints. We consider this to be an imperative component in making a strong system that assesses the megatrends recently recognized. We examined further to find how organizations endeavor to comprehend the business/society relationship. The ease and quick-changing nature of the scene rolls out this improvement hard to nail down and monitor. Many organizations react with explicit commitment of partners and following components, and we discovered a few models. It is intriguing to take note of that partner commitment isn't a panacea–there were blended perspectives on its adequacy. The organized partner commitment measure in one organization is driven by the CSR group's understanding that others' issues present business openings.

The cycle is connected to a materiality practice that covers the whole worth chain and distinguishes partners who give the best chances and who the organization needs to see better. By and large, be that as it may, commitment seems, by all accounts, to be strategic–worried about prompt issues as opposed to more generous patterns with vital importance. For sure, one interviewee remarked that partners frequently have exceptionally brief time frame skylines and are not keen on examining issues past the following several years. The commitment interaction can likewise under convey by restricting the sort of associations addressed. One member has done a considerable discussion, yet a large portion of the partners were from the business or were industry-trained professionals. Barely any friendly vested parties or customers were incorporated, gambling rather thin ends. In two cases, NGOs are unmistakably seen as having little to add to key reasoning. Albeit fundamental as accomplices in conveying explicit projects, they are viewed as having a limited viewpoint dependent on the issue they support. In that capacity, one organization trusts it is hard to separate helpful insight based on what are viewed as doomsayer messages.

Another embraces an inactive, protective stance, avoiding the chance to connect with as this would incite analysis. Others acknowledge that some commitment can be useful, yet there is a component of experimentation in discovering the gatherings that can assist organizations with understanding the evolving setting.

Acquiring key bits of knowledge is especially hard for organizations working in various nations (the greater part of those in this exploration). One organization coordinates commitment from the middle yet it is completed locally by every nation activity. Solidifying these conversations might make an abstract of nearby concerns instead of educating a wide-running vital agreement. To enhance partner inputs, a few organizations look to formal exploration and tap the assets of research organizations, including some that emphasis explicitly on CSR. One organization commissions research on expected patterns, especially in business sectors where it isn't all around addressed and consequently has less understanding than in its home market.

5.1.12. How Broad Is Mindfulness Inside the Organization?

As business procedure is the obligation of the chief initiative and the most senior administration, it is fundamental that cultural experiences acquired through the cycles we have depicted are perceived by all experts that help create or execute the system. An internal commitment measure is essential for the essential importance of any experiences to be completely perceived. The exchange between the leader group, procedure experts and CSR experts requires shared agreement. It is about the business utilizing every one of the assets available to its (the two partners across the association with various focal points for survey the world just as outside partners) to comprehend the business setting. Such shared agreement is important to recognize patterns that probably won't be the most significant on a worldwide scale yet have most pertinence for the organization's procedure–as well as the other way around, megatrends that are less pertinent in the organization's particular area and conditions. In a portion of the organizations this shared arrangement is obviously illustrated.

One methodology proficient disclosed to us that senior administration become more keen on CSR themes the more they see the connection to business achievement. One more said the climate bunch looks five years ahead, attaches its intuition to the strategies and makes mindfulness through programs that connect to business targets. A few organizations have formal board councils or chief gatherings committed to CSR-related points. Be that as it may, it is hard to pass judgment on the degree to which such bodies really draw in with vital issues.

Our examination provides reason to feel ambiguous about whether shared comprehension inside organizations is far reaching. For instance, interviewees advised us:

- As long as exercises are legitimate, ranking directors consider any to be as specialty issues.
- CSR centers around citizenship programs, instead of worldwide issues like security and tax collection.
- Some of the senior supervisory crew comprehend the significance of cultural issues, however not all.
- There is a craving among senior chiefs to take a gander at things in an unexpected way, however it is curbed.

5.1.13. Fruitful Strategizing

Characterizing fruitful procedure improvement is famously troublesome. Appropriately, our interviewees didn't offer any single answer to tackle every one of the difficulties. To be sure, some remarked that achievement can never be ensured and disappointment ought normal. Disappointment is in the idea of hazard taking–as Einstein's greatly cited saying has it: "Anybody, who never committed an attempted, had a go at nothing new." One superseding prerequisite for progress is the requirement for savvy instinct. Judgment, in view of clear qualities and convictions, is the center element for sound system. It can't counter every one of the dangers in another endeavor, however misguided thinking will subvert in any case solid cycles. The best essential reasoning won't convey achievement in the event that it prompts some unacceptable ends.

Be that as it may, while practical insight is fundamental, and could be supposed to be the substance of administrative dynamic, it is truly challenging to control or record for. The accompanying system presents judgment, close by different rules perceived in this exploration, to outline the critical components in a tough methodology and the connections between them.

1. Process Refers to the methodologies an organization has set up to foster its business technique. While we start with this component, perceive that this cycle is indispensable to how an organization makes culture and performs commitment. The cycle component envelops authoritative constructions, jobs, and obligations, and

the degree to which a business utilizes outside consultants to plan the instruments for system advancement. In pondering fitting cycles, an organization needs to respond to three key inquiries:

- Who is liable for what–should key reasoning be a cross-utilitarian or expert movement?
- How far do we depend on specialists all the while and content of system advancement?
- What is the best construction for guiding technique to move from intuition to activity?

2. Commitment Concerns the organization's inclination to construct further relations with a more extensive gathering of external consultants, specialists, and assessment formers. To be successful, this will include a skyline filtering measure that joins a more all-encompassing perspective on the setting wherein the business works. It isn't sufficiently just to comprehend future patterns according to numerous viewpoints–the sign of a genuinely tough business is one that can draw out the linkages among various (and now and then contending) perspectives on the world. Key inquiries to pose include:

- Is the organization willing to draw in with pundits of the business, just as the individuals who are steady of our exercises?
- Where are the genuine wellsprings of motivation and edification that can assist us with understanding what's to come?
- How would we guarantee the voices of specialists from nontraditional foundations both talk with power and are treated in a serious way?

5.1.13.1. Culture

Relates to the association's capacity to accept change and advance development. Somewhat, this can be affected by the cycles portrayed above–authoritative designs can help or prevent key development. Essentially, culture is emphatically affected by the judgment of the initiative group who regularly set the upsides of the business (talked about underneath). Senior administration can do a lot to advance a culture that moves individuals to think deliberately. In this regard, the chief group of an organization needs to resolve such inquiries as:

- Is our business driven by long term contemplations of benefit advancement, or transient worries of benefit augmentation?
- Is vital deduction coordinated towards hazard relief, or would we say we are worried about setting out open doors?
- Do we emphatically advance distinction and creativity among our representatives, or do we esteem and award individuals that think the same way as us?

5.1.13.2. Judgment

The last component of the essential arranging measure concerns the convictions and judgment of senior pioneers. Not at all like the other three components of this model, this center fixing is harder to control, yet it stays conceivable to impact senior administration thinking. It gets to the core of the point of view of the singular individuals from the chief group–where do they think society is going; what will be the effect of the megatrends like populace development and environmental change: what will be the job of business in the public eye in 2025? The Director of Strategy and the Chief CSR Officer can't anticipate changing these basic perspectives and decisions, however they can surface them by posing such inquiries as:

- What are the standards illuminating the chief's intuition on key choices?
- What were the purposes behind deciding NOT to seek after certain business openings?
- What are the implicit convictions of individual Executive colleagues that drive their vision of the working climate?

5.1.13.3. Reality Strategy

The interaction between procedures, culture, commitment, and judgment is what enables an organization to foster a triumphant strategy that will convey results over the long haul. Be that as it may, success can never be ensured. We generally return to the problem that Einstein alluded to–perceiving that any essential advancement carries with it the expected danger of disappointment. Be that as it may, there is a significant contrast between a fruitless procedure and helpless technique improvement. A helpless procedure with minimal likelihood of coming out on top is probably going to rise up out of a deficient methodology approach. That is a reprehensible mistake that organizations should keep away from. With regards to our exploration, it is the inability to assess cultural improvements for development and effect. An ineffective system is unique.

A technique with savvy instinct, strong cycles, and the entirety of different components portrayed above, may fall flat as a result of unforeseeable conditions, misreading of patterns, innovative vulnerabilities or different questions. Be that as it may, the system approach was the right one. The right system approach, one that outcomes in a versatile methodology–a procedure that will situate the organization for long term achievement–has a lot more noteworthy likelihood of coming out on top. A strong technique depends on a comprehension all through the organization of the potential for progress coming from the cultural setting wherein the business collaborates and works. This thusly will foster the experiences and motivation that lead to development and versatile development.

5.1.14. Reasonable Guidance on Strategy Development

We accept that to foster a strong procedure, organizations should be able to do:

- Looking comprehensively at the cultural setting to perceive interdependencies.
- Collaborating to accomplish new, productive bits of knowledge.
- Articulating the central qualities that characterize the job of business in the public arena and its commitment past the thin bounds of the accounting report.

In view of our exploration, we give these helpful bits of knowledge into making a versatile procedure:

- Corporate qualities ought to be unmistakable and significant, empowering proactive conduct and giving individuals certainty to make decisions and get things going.
- Success requires a thorough examination of a wide scope of variables.
- The business should take on a forward-looking viewpoint on the grounds that as patterns develop, the organization will be very much situated to take administration.
- Success relies upon understanding the requirements of a wide scope of partners, keeping an open exchange with various gatherings and acclimating to evolving requests.

- To install a methodology all through the organization it should be invigorating and sound appealing to the people who should get it going.
- It is important to constantly survey and adjust–presumptions might not be right however you can in any case make ventures pay by reacting quickly and fittingly to the new setting.
- An organization ought to consider the drawn out CSR of the business as a central reason for the methodology improvement measure, instead of just an empowering influence.

We add a few components for effectively coordinating cultural variables in system advancement:

- The corporate vision ought to expressly include the business/society relationship–clarifying both the business and social reason for the organization and drawing out the linkages between the two.
- Horizon checking necessities to look past the traditional business drivers to comprehend, expect, and profit by a more extensive arrangement of long term issues.
- The CSR group should be completely incorporated into the procedure advancement measure, instead of working in a vacuum without association with the corporate methodology.

5.2. BENCHMARKING

The business world is driven by information. In the economy and society, we use numbers to get what has occurred previously, and estimate what may happen later on. Speculation choices are directed by monetary information. Buyer buying decisions are educated by data on cost and quality. Indeed, even the medications that specialists can endorse have been the subject of definite calculating to set the cost of the medication and evaluate drug viability. In the realm of corporate obligation (CR) and sustainability, information is pivotal to understanding and further developing organization execution. Disclose to partners how the organization analyzes as far as top tier execution and different measures. Issues range from seeing how your organization adds to work across the worth chain to estimating the effect of local area venture programs. The cutting edge CR specialist basically couldn't manage their work without superior grade, solid information. Be that as it may, information in separation is inane. We need to contrast that information with something. Past execution in your own association sets a benchmark which is fundamental for checking progress. Yet, genuine worth stems from examinations with the more extensive world. Is your exhibition top notch? Is it true that you are falling behind the bunch of friends and best practice organizations on a specific issue? Contrasting CR and sustainability measurements and a more extensive arrangement of organizations is key to comprehension your present exhibition in setting, yet in addition your future potential. Furthermore, an organization's monetary, social, and ecological presentation should be thought about in contrast to standards, laws, and principles set by society. They go about as the external benchmark by which an organization's conduct is judged.

Benchmarking is a goal assessment, in view of clear definitions and measurements, that analyzes execution against a particular standard, standards or gathering of organizations. It is generally famous as a method for appraisal to survey an organization's exhibition with regards to friends and best-practice organizations. Inside, this can drive enhancements, make a feeling of contest and move new thoughts. For instance, if a contender has gained huge headway to dispense major struggles from minerals acquisition from its production network, they may be utilizing a clever methodology from which you could learn. Remotely, it can show partners how the organization contrasts and others and what progress it is making to resolve an issue.

In the event that your own corporate obligation technique is elite, your external acknowledgment through evaluations, norms, and grants should contrast well and direct contenders just as a more extensive arrangement of best-practice organizations. There are incalculable various sorts of benchmarking: monetary benchmarking analyzes the monetary competitiveness of organizations; measure benchmarking surveys how business methods are completed; benchmarking against a standard sets a base exhibition level to be met. Execution benchmarking evaluates how an organization oversees or resolves a specific issue in contrast with a bunch of organizations or principles. This can be quantitative, for example, contrasting the measure of water utilized with make a unit of item, or subjective, for example, looking at provider sourcing plans. The capacity to benchmark in the field of CR has been enormously helped by the expanding number, and normalization of, sustainability detailing. Around 95% of the Global 250 issue sustainability reports and associations like Global Reporting Initiative (GRI) and Sustainability Accounting Standards Board have been attempting to normalize revealed information. These standards not just communicate/

informs external partners what's going on inside the organization's sustainability procedure, yet in addition permits organizations to think about execution in a similar way within different parts of the business.

Most importantly, benchmarking permits organizations to gain from each other and settle on more astute choices. By considering execution in contrast to peers or a norm, organizations can drive gradual upgrades across the business and increment usefulness. Spearheaded by quality administration master W. Edwards Deming, ongoing exploration has shown that quality benchmarking further develops execution in fields as different as assembling measures, corporate administration, wellbeing, and security, variety, and inventory network integrity.

Critical investigation and analysis of information accumulated can likewise highlight future patterns and assist organizations with keeping up to date with and even advance beyond peers. For instance, organizations with enormous stock chains in agricultural nations progressively need to show that their crude materials and fabricated items are delivered with trustworthiness. These organizations have thought that it is basic to benchmark provider work rehearses, like the option to worker's guild participation and the end of kid and constrained work, against worldwide principles just as neighborhood laws and standards. The UN Universal Declaration of Human Rights has given a norm against which business conduct can be judged.

Benchmarking evaluations can be of specific worth to CR and sustainability leaders working in organizations. Seeing how you look at on a specific issue can advise dynamic and reinforce correspondence channels with inside and outside partners. There are two fundamental benefits for the CR professional to acquire from a powerful benchmark:

1. The case can be made that the organization needs to do more to accomplish best practice norms in a specific region. For instance, in the event that it very well may be shown that your provider set of principles is feeling the loss of a few components saw as best practice by inventory network specialists, the case can perfectly be developed that it could be an ideal opportunity to refresh your implicit rules to line up with best practice.
2. Furthermore, precise benchmarked information can back up cases of authority. This is especially valuable in cases we have seen where the organization either needs to make a case of predominance, or when execution in enduring an onslaught from outside partners. For instance, if the extent of organization

enrollment in your business is essentially higher than the public normal in each nation of activity, the information may uphold the argument against allegations of being antiorganized work.

The benchmarking has its constraints:

While great benchmarking is about execution and best practice, terrible benchmarking can prompt unremarkableness. Awful benchmarking is utilizing information to legitimize normal execution, instead of testing and driving upgrades. This typically comes from choosing some unacceptable correlation bunch. Just in light of the fact that others are performing at a specific level, doesn't mean this is the level at which you ought to perform. In the event that your whole industry doesn't resolve the issue of a living pay, this doesn't imply that your pay theory is fundamentally a decent one. As well as advocating unremarkable execution, awful benchmarking can smother advancement. While you might have a set friend correlation bunch inside your area, individual organizations and ventures dominate in various regions. Organizations that acquire the most worth from a benchmark frequently consider some fresh possibilities and hope to get thoughts from other areas by asking who is awesome at a specific issue and what are they doing another way? For instance, Southwest Airlines looked to further develop its on time execution. Rather than benchmarking against different carriers, the organization looked to NASCAR where pit groups rapidly and effectively get their race vehicles back on the track as wellspring of development.

Benchmarking can offer a new interpretation of an issue and yield basic experiences to improve execution. In any case, prior to plunging into nitty-gritty information investigation, set out what you need to accomplish. Here are five expected points (Figure 5.2):

1. **Identify a position in the peer group:** To understand overall performance, take stock of how you compare. How does my performance compare on my priority areas? Is my recent drop in performance still within typical industry practices?

2. **Inform strategy development:** To recognise areas of strength, opportunities to differentiate, key vulnerabilities and potential performance gaps. In which areas do I lag behind my peers? Where can I distinguish myself as a leader?

3. **Determine best practice elements and set targets:** To set sustainability targets and inspire action, examine the actions of best in class companies, the performance of innovative leaders, and the highest standards within an industry and across industries. How does my performance compare to world-class standards? To what degree have leading companies outside of my sector improved their performance?

4. **Understand trends over time:** To track how the market is moving on a key issue, review available data over the past years and analyse for trends. Has the variation among companies decreased? Has overall sector performance improved? Has an improvement programme been successful?

5. **Communicate performance to external stakeholders:** To build a constructive dialogue with allies, partners and even critics, make the findings from a benchmark public. A narrative should accompany any benchmark that sets performance in context.

Figure 5.2: Five steps to benchmarking.

A clear process is needed to identify the right metrics and to transform the available data into valuable insights. We recommend using a five-step process (Figure 5.3):

Figure 5.3: Five steps process to benchmarking.

The end goal of benchmarking is not a benchmark. The real benefit stems from understanding the market and identifying how to improve performance and develop new ideas. Benchmarking can have its limitations. It cannot provide all the answers.

There is often the temptation to look for a single right answer—the underlying explanation for performance differences can sometimes be complex. Most data only makes sense when understood in context, for example, where aggressive target-setting has driven superior performance for an organization. For this reason, benchmarking is best thought of as a first step. After interpreting the data, it's the action that follows that really differentiates companies. An illustrative quantitative benchmarking criteria is given in Figure 5.4.

ECONOMIC

- Spend on local procurement
- Tax contribution by country
- Direct economic value generated and distributed:
  - Employee wages and benefits
  - Tax contributions to local governments
- Ratio of annual total compensation for the highest-paid individual in each country of significant operations to the median employee salary in that country

ETHICS

- Number of ethics breaches reported to ethics hotline
- Number of incidents of non-compliance with regulations and voluntary codes around marketing
- Value and breakdown of political contributions

ENVIRONMENT

- Weight or volume of materials used
- Percentage of recycled inputs materials used to manufacture products
- Total energy consumption
- Total fuel consumption from renewable and non-renewable resources
- Total energy consumed
- Energy requirements of sold products or services
- Total volume of water withdrawn by source
- Total volume of water recycled or reused
- Tonnes of greenhouse gas emissions, included Scopes 1, 2, and 3
- Tonnes of NOx, SOx, ozone depleting substance emissions
- Volume of water discharged by quality and destination
- Breakdown of total weight of waste generated by disposal method
- Total volume of significant spills
- Number of grievances reported related to the environmental impacts

SOCIAL

- Gender breakdown of employees by employment type, management level and region
- Employee turnover by age group, gender, and region
- Percentage of managers local to the country of operations
- Value of investment in communities broken down by cash, time, and in-kind contributions
- Percentage of pre-tax profits donated
- Percentage of employees that participate in volunteering or matching gift programmes
- Lost time injury frequency rate
- Average hours of training per employee by gender and employee category
- Percentage of diverse employees by employee category
- Average salary at each employee level by gender
- Percentage of employees that are members of labour unions or covered by collective bargaining agreements
- Percentage of suppliers screened for labour issues
- Hours of human rights training per employee
- Number of reported incidents of discrimination

Figure 5.4: Illustrative quantitative benchmarking criteria.

5.3. MASTERING RESILIENT GROWTH

Key arranging by organizations needs to change. Numerous organizations are confronting intense difficulties, including how to: develop incomes in a quick evolving world; expect the up and coming age of client needs; forward-plan in a questionable economy; and keep up with discipline on costs. However, such a large number of organizations take on a foolish and restricted way to deal with key administration. They center around beating their rivals by coordinating with their exercises, creating comparable items through steady upgrades and conveying esteem through the normal, worn out plans of action. They need to utilize the old models however make them work more enthusiastically. The drawn out achievement of organizations today is at this point not ensured. To be sure, the actual presence of numerous organizations is under danger. The normal life expectancy of a S&P 500 organization has contracted from 61 years in 1958 to only 18 years today.

History is full of companies that failed because they neglected to adjust sufficiently quick to new market elements, like Blockbuster video, Pan American Airlines and Polaroid cameras. An extreme reexamining is required. The way in to what's to come is getting what we call the unique circumstance–the patterns, effects, and connections between a business and the more extensive world. Examining the setting of any business can uncover incredible bits of knowledge for advancement. This can support execution as long as possible. A tough procedure is one that is in sync with an organization's more extensive setting, permitting the business to flourish for quite a long time to come.

Business doesn't exist in a vacuum. Setting alludes to the financial, social, political, mechanical, and natural issues that set out business open doors and dangers. These variables have customarily been considered as a feature of a restricted danger evaluation by most organizations. In any case, there is a greater chance. An examination of applicable connections, effects, and patterns can give new bits of knowledge. When drawn closer deliberately and comprehensively, Corporate Citizenship's Resilient Growth approach can assist with enduring the hardships and reveal the chances from startling guidelines, changing client designs and more extensive financial variances. This section sets out how any ground breaking business can comprehend its specific circumstance, examine the opportunities, open new experiences and advance for long term development. It provides insights into the tools that businesses can use to foster an effective and coherent methodology in a fast-changing world.

5.3.1. Planning the Unique Situation

No business is an island. Imperceptible bonds shape the firm through a complicated communication of financial, social, political, mechanical, and natural patterns. Organizations that need to be situated for long term achievement need to acquire a comprehensive comprehension of the connections that shape their unique circumstance. This should comprise of:

1. Understanding the organization's mind boggling connections in its many business sectors, and with its numerous partners.
2. Dissecting the sorts of market influences and impacts that can possibly affect business achievement today and later on.
3. Distinguishing how to coordinate this speculation into business procedure and intending to catch the most esteem and guarantee long-term achievement.

There is a two-way communication between the business and its unique situation. Society makes expenses and an incentive for organizations and business makes worth and expenses for society. A more profound examination of these connections can give a springboard to advancement.

5.3.2. Future Change

The setting is dynamic, not static. A billion new buyers are added to the world each fifteen years. Around 4.2 billion individuals utilize a cell phone to get to web-based media today. Some progressions can be astonishing. Who might have conjectured in 1990 that interest for natural food would detonate by 3,000% to turn into a $30 billion market–4.2% of all food deals in the United States today? Regulation can be emotional as well: in 2008, China restricted super flimsy plastic packs while Denmark presented an assessment on food sources high in soaked fat. A few organizations react by keeping the principles. This might be vital on account of laws, and alluring on account of standards, however it isn't adequate to acquire the upper hand. Others go further and look to recognize the chances for "mutual benefits" where they can profit from a change. These covering interests can make examples of overcoming adversity in select regions. The most nimble organizations embrace change and use it for their potential benefit. Dominating Resilient Growth is a foundational reaction: expecting change, appreciating the associations, bridling new bits of knowledge, molding the actual setting and making genuine upper hand that can be supported for quite a long time to come.

5.3.3. Advancement with a Wide-Point Focal Point

Advancement is fundamental to open the chances for Resilient Growth. The business setting ought to be surveyed comprehensively with a 'wide point focal point's that ganders at an expansive scope of elements throughout different time scales, including 10, 20, and 30 years ahead. By extending this edge of vision, organizations can recognize suffering freedoms that can be supported for long term achievement. Breaking down the setting can reveal insight into new plots for organizations to improve all the more inventively, diminish dangers and fabricate more prominent flexibility. At its heart, this interaction is tied in with recognizing and reacting to beforehand unidentified, implied or neglected necessities through new items, administrations, and plans of action. These regularly come from an examination of patterns and effects, and how the interdependencies between them could shape startling results later on.

5.3.4. What Sorts of Progress Can Arise?

- New items and administrations that meet a formerly neglected need (for example minimal expense energy; safe food and drinking water; further developed schooling and wellbeing; upgraded foundations and framework)
- New plans of action that further develop efficiencies and yet again use items in a shut circle model that makes additional worth from less sources of info
- Employment and occupation creation in the store network that might include disregarded gatherings (for example smallholder ranchers; ladies; the jobless)
- New proprietorship structures that offer rewards in an unexpected way (for example cooperatives)
- New coordinated efforts with providers, joint endeavor accomplices, clients, and non-benefits that meet common objectives

5.3.5. What Are the Possible Business Benefits?

- Top line development from new item and administration advancement
- Lower cost of merchandise from further developed efficiencies
- Faster reaction to new guidelines than contenders
- Improved discernibility in the production network
- Reduced dangers and expanded flexibility of business tasks
- Improved wellbeing and security record
- Reduced exchange costs from more significant levels of trust (for example with clients)

- Enhanced standing and brand value
- Improved worker inspiration and efficiency
- Increased innovativeness and thought age
- Attraction of better ability and decreased staff turnover
- New collusions and more beneficial coordinated efforts through shared objectives
- Lower cost of capital from a lower hazard profile

The chances are interesting to each organization. They will rely upon the particular setting and friends aspirations. No two cooperations are something similar and these issues change from one organization to another, just as landmass to mainland. The unique idea of the setting implies that new issues are continually arising. Forward-looking organizations should try to catch the rising worth before contenders do. However, driving organizations don't simply respond to patterns; they effectively attempt to shape them for their potential benefit. Organizations needing to reclassify their business sectors don't simply acknowledge the future, they try to change the standards of the game and change frameworks and individuals for their potential benefit.

5.3.6. Long Term Value Optimization

Business adventures should be beneficial over the long haul. Canny business pioneers realize that this doesn't generally mean the goal ought to be momentary benefit augmentation. There are no proper standards about the timescale over which worth should be made. It is society's pressing factors–from nearsighted predispositions to securities exchange assumptions–that regularly oblige organizations. Throughout the next few years, we hope for something else and more organizations to discuss a shift away from momentary benefit amplification to long term esteem advancement. Tough Growth recognizes new advertisement helps that can support benefits and investor esteem over the long haul.

5.3.7. How Resilient Growth Makes An Incentive for Organizations

A few organizations are now exploiting these kinds of bits of knowledge. An anecdotal contextual analysis shows how the cycles can be applied. It depends on our experience of working with organizations on these issues. ABC Pharmaceuticals has experienced a lack of development, declining net revenues and a pipeline which financial backers see as unsuitable. Representative confidence is low, innovativeness is missing and a progression of issues have escalated into adverse media remarks. An essential audit utilizing the Mastering Resilient Growth approach would start by understanding the association's most noteworthy impacts on society and how value is created for the business from society.

Likewise, benchmarking recognizes a bunch of new medium to long term dangers and openings. From an extensive rundown of likely thoughts, another knowledge is distinguished to offer a groundbreaking wholesome item at a lower cost to a lot more individuals. This significant change has so far been undiscovered. A creative new plan of action is needed to understand the aspiration and guarantee it conveys inside asset and spending limitations. Close by offering the food, the organization gives a free nearby wellbeing check to buyers. An association with limited scope merchants empowers the help to make back the initial investment and arrive at scale all the more rapidly. This opens up new courses to advertise that were beforehand undiscovered, and permits more items to be sold. In the interim, functional, and store network capacities set out on a program of improved detectability for the item and a more slender creation measure. This conveys a lower cost of merchandise, a greater item and a stronger inventory network. Following quite a while, the endeavor is productive after fast volume development that was unrealistic under the old plan of action. Close by top-line development and rising net revenues, new ability is drawn to the organization. Worker assurance and imagination rise. As a feature of the business turnaround, the more drawn out term pipeline improves. The financial exchange invites the outcomes and an expanded trust in administration is confirmed through a lower cost of capital. More grounded associations with partners–like clients, controllers, and the media–make a more grounded brand, more elevated levels of trust and support rising overall revenues into what's to come. Obviously, this anecdotal model sounds unrealistic. However, it's not a long way from the kinds of new items that organizations like Danone (Shakti Doi yogurt), Vodafone (M-Pesa cash move), Unilever (Pureit water purifier) and Cemex (moderate lodging in Mexico) have brought to showcase. Truly, the excursion is rarely direct and there are set-backs en route. In any case, the proof shows that these kinds of approaches are conceivable and can be beneficial.

5.3.8. Methods of Working As Long As Possible

Powerful system expects organizations to get and support the upper hand. Tough Growth is a perspective that can open worth through taking a gander at the world recently. In the midst of progress, it can give the new bits of knowledge expected to take off another contender, increment incomes and support development into what's to come. It gives another dialect and new speculation for business. Two better approaches for believing are needed to completely understand the advantages of Resilient Growth. Right off the bat, interdependencies should be truly perceived. No business is an island, yet in addition no issue exists all alone. Patterns in the setting interface and shape each other. How should purchaser assumptions shape guideline? How is it possible that a new innovation would impact the inventory network? The most beneficial experiences frequently come from looking comprehensively, not at single issues in seclusion. Also, better approaches for working through coordinated effort can achieve greater advantages for all. Coordinated effort starts through conversation with the rest of the world at the Insights stage.

Organizations frequently don't concoct extraordinary thoughts all alone. Association can likewise stretch out to connecting up with the right provider, joint endeavor accomplice or non-benefit at the execution stage. This is on the grounds that viable organization can give new bits of knowledge, decrease expenses for the business, scale a drive all the more quickly and increment the size of the pie for all. Applying the thoughts by and by is no simple undertaking. Dynamic requires information however not all things can be precisely estimated. Directors should be open to managing vulnerability. Compromises might be required. No arrangement can be awesome and here and there five stages forward requires three back. This implies that the director's outlook might have to change for Resilient Growth methodologies to be viable. Strong Growth implies an associated and commonly useful connection between the organization and society. The prize is a business that is more inventive and ground breaking than the opposition; a firm that expects new client needs, and encounters less astonishments on the lookout. However, in particular, Resilient Growth guarantees that the organization can support its exhibition a seemingly endless amount of many years. By calibrating the business to be in a state of harmony with the rest of the world, the organization can flourish for quite a long time to come.

5.3.9. The Glad Side-Effects of a Tough Technique

By being more on top of the world and its assumptions, organizations advantage their partners, society, and the climate en route. Business can be an incredible power for social great by applying their assets deliberately in order to guarantee long-term productivity that is in sync with society. We find that "being seen" to be "doing acceptable" just for reputational reasons is normally less productive than distinguishing the smart activities that can assist the business with prevailing in the long haul. We tell our customers that social advantages and more grounded notorieties ought not be considered as the objectives of such a methodology, however a supporting side-act, the cheerful result of a supported spotlight on business achievement.

5.4. ENHANCING BOARD COMMITMENT

Corporate sustainability is best when sponsored by convincing help from senior decision-makers. Capable strategic approaches are as of now not the protection of one office. They should slice across issues and groups to acknowledge genuine incentive for the business and society. While many organizations guarantee that sustainability is significant, scarcely any sheets methodically think about the issues. The way to conquer this hole is to interpret corporate obligation from a specialty, specialized issue into a language that resounds with sheets. That implies clarifying how mindful strategic approaches support the drawn out benefit of the firm. Directors looking to get purchase in for sustainability need to comprehend the center abilities of the board. Viewpoints; abilities and information shift, and organizations wind up at various phases of the sustainability venture. A bunch of nine ways to deal with welcoming the board 'on board' can help slice through the commotion and improve on sustainability for senior leaders.

Powerful corporate sustainability requires great corporate administration. Firms should guarantee that senior chiefs comprehend and can react to the difficulties and chances of economic turn of events. This is especially obvious today in light of the fact that corporate sustainability traverses assorted issues, from new item developments to burden arrangements. However, sustainability directors every now and again experience hindrances to connecting with leaders. This paper sees how to comprehend the board and their differing skills. It recognizes the most proper devices to get purchase in to sustainability drives. It is planned for public organizations with a Board of Directors, yet the experiences will apply similarly to various degrees of the executives and different sorts of association Responsible strategic approaches have developed in ongoing a long time from impromptu, altruistic action to key, expansive drives that adjust the social, natural, and financial effects of a firm with business achievement. Chiefs entrusted with executing these projects should guarantee senior-level purchase in to concur the degree of desire; arrange drives across various specialty units; and convey programs that advantage the entire organization. To understand the benefits for the firm and society, switches should be pulled from the top. An examination ready for the World Economic Forum tracked down that empowering sheets to more readily comprehend the all out ecological effects of an organization's items and administrations would be the single most prominent supporter of a stage change in sustainability around the world.

Despite the fact that CEOs and friends chiefs say that sustainability is a need, not many sheets give the issues a lot of unmistakable quality. All things being equal, the board's essential consideration frequently just spotlights on a narrowly defined set of monetary controls. Non-monetary effects get passing notice–or neglect to make the plan by any means. Yet, there is a reasonable acknowledgment of a requirement for the more noteworthy obligation to sustainability from organization chiefs. As indicated by a United Nations study, 93% of CEOs say that sustainability is essential to their organization's future success. The Companies Act in the UK expresses that chiefs should have "respect to the effect of the organization's procedure on the local area and the climate." Sarbanes-Oxley in the United States forces to distinguish and unveil material issues. However, dynamic board commitment in dependable strategic approaches remains low. Indeed, when gotten some information about their needs for the year ahead, only 2% of organization chiefs everywhere American firms refer to "CSR." All things being equal, "key arranging" positions highest. The clear hole among "CSR" and "vital arranging" is the key motivation behind why scarcely any sheets offer unmistakable quality to corporate sustainability. To defeat this, administrators need to break corporate obligation out of its storehouse and clarify it in a language that reverberates with the board. Great corporate administration expects chiefs to have the foreknowledge to guide an organization through unchartered waters.

That implies an oversight of: long term system; changing business sector elements; hazard controls; financial backer viewpoints; administrative consistency; and notoriety the board. These issues are central to present day corporate sustainability. Large numbers of our customers have revealed to us that their greatest boundary to inside purchase in is getting time on the load up's bustling plan. Be that as it may, time is just a large portion of the story–what issues is the load up being approached to consider? While some may turn off at the notice of energy the executives or worker chipping in, similar individuals sit up while talking about the drawn out dangers and openings confronting the organization. Mindful strategic approaches ought to be converted into a language that presents them as deliberately important to the drawn out benefit of the firm. A bunch of instruments to assist with articulating the connection between sustainability and business achievement are researched later. Prior to drawing in with board chiefs, comprehend a singular's beginning stage. Not all board chiefs are something similar. The basic capabilities of the board can be separated into points of view, abilities, and information. It is fundamental to comprehend these distinctions to adequately connect with the two people, on a one-to one premise, just as a board aggregately. Points of view are the mentalities and upsides of people. They identify with good and close to home needs, like genuineness, uprightness, esteem decisions, encounters, and even predispositions. Viewpoints are profound situated individual characteristics that advance gradually over the long run. Points of view are hard to change, however a few chiefs depict an 'revelation second' that has done precisely that.

Abilities contrast according to points of view in that they are values-unbiased and useful. They incorporate aptitude, for example, in partner relations and organization working, just as bits of knowledge, like the capacity to think about various perspectives. In contrast to points of view, abilities are agreeable to being instructed and created. Information alludes to the data that is accessible for dynamic. Handy loads up need to approach the right data, at the ideal opportunity, to settle on compelling choices. Information might come from related knowledge (which influences viewpoints), or another source. Streams of information can be expanded. For instance, board individuals can be advised on new information on non-monetary issues, like changing cultural assumptions, natural freedoms, representative inspirations and long term industry elements (Figure 5.5).

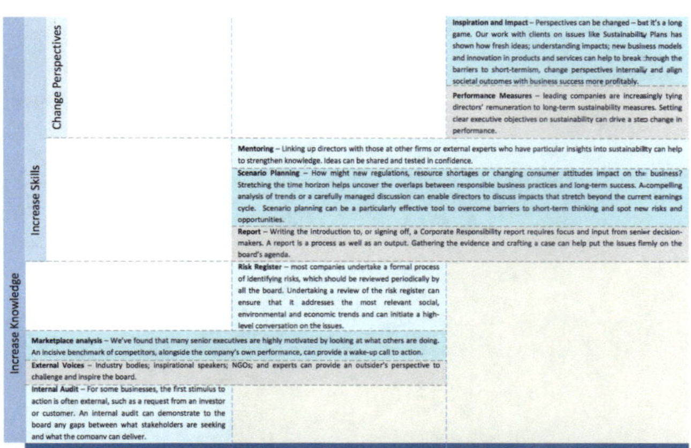

Figure 5.5: Board engagement strategies.

For the sustainability expert, the central questions to consider are:

1. Where is your board on the sustainability venture?
2. Except if you are extremely near the CEO, and exceptionally persuasive, you are probably not going to have the option to change their profound situated points of view. Yet, you can open the board to new data and urge them to foster new abilities.
3. Give motivation and groundbreaking thoughts.

New data sources are expected to spruce up an organization's way of dealing with corporate sustainability and assist chiefs with recognizing the advantages. Understanding these three elements of decision-making will assist you with choosing the proper methodology to draw in the board. The board ought to be asked to think about an inquiry: where does sustainability "fit" on the organization's plan in the short and long haul? Is it part of the corporate vision? Or then again lined up with inventory network and hazard work? Explaining the degree of desire, just as the space that sustainability sits inside, is fundamental to viably following through on a program.

Without this knowledge, progress will be testing. Putting sustainability at the core of senior dynamic is a drawn out game. It requires better approaches for deduction, new thoughts and approaches that are custom-fitted to every interesting setting. Successful commitment of the directorate should prompt a more grounded obligation to corporate sustainability. However, the genuine prize will come from an exhibition to the board that, in the long haul, business achievement can go inseparably with more prominent advantages to society.

CHAPTER 6

CLIMATE CHANGE

CONTENTS

6.1. WHAT IS CLIMATE CHANGE?

1. Climate change can be a characteristic interaction where temperature, precipitation, wind, and different components fluctuate over many years or more. In a huge number of years, our reality has been hotter and colder than it is currently. However, today we are encountering uncommon quick warming from human exercises, basically because of consuming petroleum derivatives that produce ozone harming substance discharges.
2. Increasing ozone-depleting substance discharges from human movement carry on like a sweeping folded over the earth, catching the sun's intensity and raising temperatures.
3. Examples of ozone harming substance discharges that are causing environmental change incorporate carbon dioxide and methane. These come from consuming petroleum products like fuel for driving a vehicle or coal for warming a structure. Clearing area and woodlands can likewise deliver carbon dioxide. Landfills for trash are another source. Energy, industry, agribusiness, and garbage removal are among the significant producers.
4. Greenhouse gas fixations are at their most significant levels in 2 million years and keep on rising. Subsequently, the earth is around 1.1°C hotter than it was during the 1800s. The last ten years were the hottest on record.
5. Many individuals think environmental change principally implies hotter temperatures. However, temperature increase is just the start of the story. Since the Earth is a framework, where everything is associated, changes in a single region can impact changes in all others. The outcomes of environmental change presently incorporate, among others, serious dry spells, water shortage, extreme flames, rising ocean levels, flooding, dissolving polar ice, horrendous tempests and declining biodiversity.
6. People are encountering environmental change in different ways. It influences our wellbeing, capacity to develop food, lodging, security, and work. A few of us are now more helpless against environmental influences, for example, individuals living in little island creating States. Conditions like ocean level ascent and saltwater interruption have progressed to where entire networks have needed to move. Later on, the quantity of "environment evacuees" is supposed to rise.

7. Every expansion in a worldwide temperature alteration matters. In a 2018 report, a great many researchers and government commentators concurred that restricting worldwide temperature increase to something like 1.5°C would assist us with keeping away from the most terrible environment influences and keep a bearable environment. However, the ongoing way of carbon dioxide discharges could increment worldwide temperature by as much as 4.4°C before the century's over.
8. The outflows that cause environmental change come from all aspects of the world and influence everybody, except certain nations produce substantially more than others. The 100 least-producing nations create 3% of complete discharges. The 10 biggest producers contribute 68%. Everybody should make an environment move, yet individuals and nations making a greater amount of the issue have a more noteworthy obligation to act first.
9. Climate change is a gigantic test, however we definitely know numerous arrangements. These can convey financial advantages while working on our lives and safeguarding the climate. We additionally have worldwide arrangements to direct advance, for example, the UN Framework Convention on Climate Change (UNFCCC) and the Paris Agreement. Three general classifications of activity are: cut emanations, adjust to environment effects and money required changes.
10. Switching energy frameworks from petroleum derivatives to renewables like sunlight based will lessen the discharges driving environmental change. However, we need to begin at this moment. While a developing alliance of nations is resolving to net no emanations by 2050, about a portion of discharges cuts should be set up by 2030 to continue to warm underneath 1.5°C. Non-renewable energy source creation should decline by around 6% each year somewhere in the range of 2020 and 2030.
11. Adapting to environment results safeguards individuals, homes, organizations, livelihoods, framework, and normal biological systems. It covers current effects and those possible later on. Transformation will be required all over the place, however should be focused on now for the most weak individuals with the least assets to adapt to environment perils. The pace of return

can be high. Early admonition frameworks for catastrophes, for example, save lives and property, and can convey benefits up to multiple times the underlying expense.

12. We can take care of the bill now, or pay the consequences later on. Environment activity requires critical monetary speculations by state run administrations and organizations. In any case, environment inaction is immeasurably more costly. One basic advance is for industrialized nations to satisfy their obligation to give $100 billion every year to emerging nations so they can adjust and move towards greener economies.

6.2. PHYSICAL SCIENCE

1. Human exercises have warmed the air, sea, and land, creating far and wide and quick changes in the climate, sea, cryosphere, and biosphere.
2. The size of ongoing changes across the environment framework is remarkable over numerous hundreds of years to a long time. Many changes are irreversible for hundreds of years to centuries, particularly as far as the sea, ice sheets and worldwide ocean level.
3. Human-initiated environmental change influences each locale. There is developing proof of connections to outrageous heatwaves, weighty precipitation, dry seasons and hurricanes.
4. Global surface temperature will keep on expanding until essentially the center of the 100 years. Except if we make sharp decreases in ozone depleting substance outflows in coming many years, an Earth-wide temperature boost will surpass 1.5°C, after which environment outcomes will be significantly more extreme.
5. The more the world warms, the more noteworthy the progressions in the environment framework become. This incorporates more continuous and serious hot limits, marine heatwaves (MHWs), weighty precipitation, agrarian, and biological dry spells in certain districts, the extent of extreme hurricanes, and decreases in Arctic ocean ice, snow cover and permafrost.
6. Continued an Earth-wide temperature boost will additionally escalate the worldwide water cycle, making it more factor, and changing rainstorm precipitation and the seriousness of wet and dry occasions.

7. As carbon dioxide outflows rise, the sea and land will be less powerful at retaining and easing back the aggregation of carbon dioxide in the climate.
8. With further a worldwide temperature alteration, each district will progressively encounter changes in the drivers of climatic effects. Drivers will be more far and wide at 2°C contrasted with 1.5°C and surprisingly, more so at more significant levels of warming.
9. Ice sheet breakdown, sudden sea flow changes and warming past momentum projections are more uncertain results yet can't be precluded.
10. Limiting human-initiated an unnatural weather change requires restricting aggregate carbon dioxide emanations, coming to essentially net zero. Solid decreases in other ozone harming substance outflows, for example, methane would likewise be required.
11. Achieving low or extremely low ozone-depleting substance discharges would lead inside the space of years to perceptible impacts on ozone harming substance and spray fixations and air quality. Detectable contrasts in worldwide surface temperature would arise in close to 20 years.

6.3. TEMPERATURE RISE

1. The Earth is currently around 1.1°C hotter than it was during the 1800s. We are not on target to meet the Paris Agreement focus to hold worldwide temperature back from surpassing 1.5°C above pre-modern levels. That is viewed as far as possible to stay away from the most horrendously terrible aftermath from environmental change.
2. 2015–2019 were the five hottest years on record, while 2010–2019 was the hottest ten years on record.
3. Global surface temperature has expanded quicker starting around 1970 than in some other 50-year time frame over essentially the most recent 2000 years.
4. On the ongoing way of carbon dioxide discharges, temperature could increment by as much as 4.4°C before the century's over.
5. In 2019, ozone harming substance focuses arrived at new highs. Carbon dioxide levels were 148% of preindustrial levels.

6. Greenhouse gas fixations, as of now at their most significant levels in 2 million years, have kept on rising.
7. Since the mid-1980s, Arctic surface air temperatures have warmed somewhere around two times as quick as the worldwide normal, while ocean ice, the Greenland ice sheet, and icy masses have declined over a similar period and permafrost temperatures have expanded.
8. Emissions should drop 7.6% each year from 2020 to 2030 to hold temperatures back from surpassing 1.5°C and 2.7% each year to remain underneath 2°C.
9. The outflows hole in 2030, or the distinction between vital carbon dioxide decrease and latest things, is assessed at 12–15 gigatons carbon dioxide same (Gt CO_2e) to restrict an Earth-wide temperature boost to underneath 2°C. For the 1.5°C objective, the hole is 29–32 Gt CO_2e, generally comparable to the joined outflows of the six biggest producers.
10. To follow a 1.5°C-reliable pathway, the world should diminish non-renewable energy source creation by around 6% each year somewhere in the range of 2020 and 2030. Nations are rather arranging and extending a typical yearly increment of 2%, which by 2030 would bring about over two times the creation reliable with as far as possible.

6.4. ADAPTATION

1. Adaptation to environmental change shields individuals from higher temperatures, increasing oceans, fiercer tempests, eccentric precipitation and more acidic seas. Certain individuals are more powerless against these impacts, like those living in destitution.
2. Small island creating States are especially helpless without transformation to tempests and ocean level ascent. For a portion of these nations, debacle related financial misfortunes have proactively been basically as high as 200% of the size of a public economy.
3. Estimated yearly transformation costs in agricultural nations are in the scope of $70 billion, however could reach $300 billion by 2030. Only 21% of global environment finance goes to variation and flexibility, about $16.8 billion every year.

4. Globally, a $1.8 trillion interest in early advance notice frameworks, environment versatile foundation, improved dryland agribusiness, worldwide mangrove insurance and strong water assets could create $7.1 trillion in kept away from expenses and social and natural advantages.
5. Over 60% of nations have nature-based systems in public environment activity designs; a comparative offer has recognized that variation relies upon safeguarding environments and biodiversity.
6. Better climate information alongside early advance notice and crisis the executives frameworks decrease actual harm and financial misfortunes. Widespread admittance to early advance notice frameworks can convey benefits up to multiple times the underlying expense.
7. Without versatile measures, the quantity of individuals who need adequate water for no less than one month out of each year will take off from 3.6 billion today to in excess of 5 billion by 2050.
8. Solar-controlled water system, weather conditions ready frameworks, new harvest assortments and other versatile measures can assist with staying away from a drop-off in worldwide farming yields by up to 30% by 2050.
9. Improving wellbeing frameworks could assist with forestalling 250,000 extra environment related passings each year from 2030 to 2050, fundamentally from avoidable causes, for example, lack of healthy sustenance, jungle fever, looseness of the bowels and intensity stress.
10. Worldwide, only 38% of little firms have put resources into adjusting to natural dangers, contrasted with 60% of huge firms. During the COVID-19 pandemic, organizations ready to endure interruption were multiple times less inclined to lay off representatives and bound to have stable deals.

6.5. FINANCE

1. Public environment finance, including $100 billion that created nations have consented to give to emerging nations every year, upholds the basic foundation for variation, strength, and the new sustainable power based economy.

2. Based on the latest evaluation led by created nations, all out environment finance figuring in with the $100 billion responsibility keeps on missing the mark, coming to $78.9 billion out of 2018.
3. Although transformation finance expanded all the more quickly between 2016–2018, its general portion of complete public money was just 21% in 2020. Transformation costs for agricultural nations might be in the scope of $140 billion to $300 billion every year by 2030, and $280 billion to $500 billion every year by 2050.
4. Private money could give the greatest pool of capital. Assuming adequate worldwide public environment finance is conveyed to prepare private assets, it will be feasible to move from the billions to the trillions required.
5. Over 160 firms with $70 trillion in resources have worked together to control the worldwide economy towards net-zero emanations and convey Paris Agreement objectives.
6. The consolidated worldwide financial reaction to the pandemic was $18 trillion as of March 2021. A similar conclusiveness is basic in the reaction to the environment emergency.
7. COVID-19 recuperation bundles have not been green, overall, in spite of overall population support for a green recuperation. In the Group of 20 nations, $250 billion has been coordinated to petroleum derivatives rather than just $146 billion to clean energy.
8. The extreme financial effects of COVID-19 cutoff the capacity of many agricultural nations to put resources into recuperation as well as environment activity. The pandemic pushed half of low-pay and least formed nations into obligation pain or high risk of it. While some obligation help is accessible for 34 nations in danger of default, 9 nations are not qualified. They incorporate little island creating States with intense environment weaknesses.
9. Needed foundation venture is around $90 trillion by 2030; the new framework should be viable with environmental objectives. Putting resources into the strong foundation in non-industrial nations could convey $4.2 trillion over the lifetime of the new framework. By and large, yields $4 in benefits.
10. Switching to a perfect economy could raise $2.8 trillion through carbon cost incomes and the redirection of petroleum product endowments to public speculations.

6.6. GENDER EQUALITY

1. Empowering ladies and young ladies to have a voice and a job in decision-production on environmental change-related issues is fundamental for supportable turn of events and more prominent orientation fairness.
2. As early adopters of numerous new agrarian procedures, first responders in quite a while, business people of environmentally friendly power energy and leaders at home, ladies offer important bits of knowledge and arrangements into better dealing with the environment and its dangers.
3. The environment emergency isn't "sexually impartial." It fuels existing imbalances, passing on ladies and young ladies to encounter novel dangers to jobs, wellbeing, and security.
4. Women are less ready to face environmental change because of restricted admittance to and control of ecological labor and products, less support in navigation and the conveyance of natural administration benefits.
5. During times of dry spell and whimsical precipitation, numerous ladies in low-and lower-center pay nations, who rely upon agribusiness, work harder to get pay and assets for their families.
6. Current environment finance seldom arrives at ladies and their associations, and just a little extent of subsidizing centers around their necessities, privileges, and arrangements.
7. As environmental change drives struggle across the world, ladies, and young ladies face expanded weaknesses to all types of orientation based brutality.
8. When calamities strike, ladies are less inclined to get by and bound to be harmed because of restricted admittance to data, portability, independent direction, as well as assets and preparing.
9. Women's and young ladies' wellbeing is imperiled by environmental change and calamities by restricting admittance to administrations and medical care, as well as expanding risks connected with maternal and kid wellbeing.
10. Extreme intensity builds occurrence of stillbirth. Environmental change likewise builds the spread of vector-borne sicknesses like intestinal sickness, dengue fever, and Zika infection, which are connected to dangerous maternal and neonatal results.

6.7. ECONOMY

1. Climate activity isn't a financial plan buster or economy-wrecker. Moving to a green economy could yield a direct monetary increase of $26 trillion through 2030 contrasted and the same old thing. This could create more than 65 million amazing failure carbon occupations.
2. Significant interest in framework is required over the course of the following 15 years, around $90 trillion by 2030. New foundation should be viable with environmental objectives.
3. Investing in tough foundation in non-industrial nations could convey $4.2 trillion over its lifetime. Overall, yields $4 in benefits.
4. More smaller, associated, and composed urban areas are worth up to $17 trillion in monetary reserve funds by 2050 and will invigorate financial development by further developing admittance to occupations and lodging.
5. Sustainable agribusiness and solid woods assurance could produce more than $2 trillion every time of monetary advantages, make a large number of occupations and further develop food security while conveying over 33% of the environmental change arrangement.
6. Doubling worldwide environmentally friendly power limit by 2030 could save the worldwide economy somewhere in the range of $1.2 and $4.2 trillion every year, to a great extent because of a monstrous decrease in costs from contamination.
7. Putting a cost on carbon and gradually getting rid of petroleum product endowments could raise $2.8 trillion that could be reinvested in open needs.
8. In 2020, G20 state run administrations committed $233 billion to exercises that help petroleum derivative creation and utilization, contrasted with $146 billion with sustainable power, energy proficiency and low-carbon choices like cycling and walker frameworks.
9. Better water the executives could further develop monetary development rates in certain districts by up to 6%.
10. The expenses of adjusting to environmental change in creating economies might really depend on $300 billion by 2030. Be that as it may, putting resources into flexibility might reduce post-calamity mediation expenses considerably.

6.8. JOBS

1. A green progress, including a shift to sustainable power, the assembling of EV, and the development of energy-effective structures, will make 24 million positions by 2030, undeniably beyond what the 6 million that could be lost.
2. By 2030, among 163 financial areas, simply 14 are anticipated to have business misfortunes of in excess of 10,000 positions around the world, and just two, refining, and separating oil, show misfortunes of 1 million or more positions.
3. If urban areas in 21 developing business sectors focus on environment brilliant development in their recuperation plans, they stand to acquire as much as $7 trillion in speculations and could make 144 million new positions by 2030.
4. Under certain circumstances, occupations made by the sustainable power industry could add up to 42 million overall by 2050–all that anyone could need to balance positions lost in petroleum product enterprises, with additional individuals ready to track down work in assembling, introducing, working, and keeping up with environmentally friendly power frameworks.
5. Jobs in sustainable power came to 11.5 million around the world in 2019.
6. Heat pressure could diminish absolute working hours overall by 2.2%–an efficiency misfortune comparable to 80 million everyday positions—and could cut worldwide GDP by $2.4 trillion out of 2030.
7. A roundabout economy, in view of the standards of lessen, reuse, and reuse, could make around 6 million new positions in reusing and squander the executives.
8. Solar photovoltaic businesses made a few 3.8 million positions in 2019. Other huge generators of new positions in environmentally friendly power were biofuels at 2.5 million positions, hydropower at near 2 million positions and wind at 1.2 million positions.
9. Jobs in renewables are more orientation adjusted than in the more extensive energy field, with ladies holding 32% of the absolute in 2019. In petroleum derivatives, they have just 21%.

6.9. ENERGY

1. Despite the COVID-19 pandemic, a record 260 gigawatts of environmentally friendly power limit was added universally in 2020, beating the past record by right around 50%.
2. Renewables became right around 5% each year somewhere in the range of 2009 and 2019, outperforming non-renewable energy sources at 1.7%.
3. Although most new sustainable power limit establishments have been made in agricultural nations over the most recent two years, created nations had multiple times more limit per capita than emerging nations in 2019.
4. In 2018, the portion of environmentally friendly power in absolute energy utilization added up to 17.1%, with the biggest expansion in the portion of renewables for power. The vehicle and warming areas show a lot more slow or no advancement.
5. More than 80% of all new power limit added in 2020 was sustainable with sun based and wind representing 91%. Interest in seaward wind hit its most elevated level ever at $29.9 billion.
6. In 2018, global public monetary streams to non-industrial nations on the side of clean energy added up to $14 billion, a 35% decline from an unsurpassed high of $21.9 billion of every 2017.
7. To cutoff worldwide temperature climb to 1.5°C in accordance with the Paris Agreement, energy progress speculation should increment by 30% for a sum of $131 trillion by 2050, yet will yield a total recompense of something like $61 trillion by 2050.
8. Global interest in inexhaustible power limit totaled $303.5 billion out of 2020, a 2% increment from 2019. However, to arrive at worldwide environment objectives, yearly interest in renewables must no less than triple by 2030, for an absolute increment of 200%.
9. Due to the COVID-19 pandemic, worldwide energy utilize fell by 4% in 2020, and carbon dioxide emanations declined by just about 6% in 2020. However, outflows have gotten back to their upwards direction and in December 2020 were around 2% higher than in 2019, preceding the pandemic.
10. More than half of the inexhaustible limit included 2019 accomplished lower power costs than new coal. New sun

powered and wind projects are undermining the least expensive existing coal-terminated plants. Sun based photovoltaics showed the most honed cost decline more than 2010–2019 at 82%, trailed by gathered sun oriented power at 47%.

6.10. FOOD AND AGRICULTURE

Delivering food causes emanations of carbon dioxide, methane, and other ozone depleting substances in different ways. Deforestation and getting free from land for horticulture and eating, assimilation by cows and sheep, and the creation and utilization of composts and fertilizer for developing yields all-cause discharges, as does the utilization of energy to run ranch hardware or fishing boats, generally with petroleum derivatives. Emanations additionally come from bundling and conveying food.

1. The unreasonable utilization of land, soil, water, and energy for food adds to ozone depleting substance emanations that cause increasing temperatures. Higher temperatures thus influence assets to create food. Up to 811 million individuals on the planet confronted hunger in 2020, upwards of 161 million a greater number of than in 2019.
2. Systems to create, bundle, and disperse food produce 33% of ozone depleting substance discharges and cause up to 80% of biodiversity misfortune. Without mediation, food framework outflows will probably increment by up to 40% by 2050, given rising interest from the populace, more pay, and dietary changes.
3. The food framework presently represents around 30% of the world's complete utilization of energy, most still delivered utilizing petroleum derivatives that create emanations.
4. Over 17% of food is squandered, and up to 10% of worldwide ozone harming substance discharges are related with food that isn't eaten.
5. Under higher temperatures, decreases in crop yields are possible. Heat pressure additionally brings about impeded quality and expanded squander.
6. The sea has consumed more than 90% of the overabundance heat in the environment framework, making it more acidic and less useful. This alongside practices, for example, overfishing

undermines marine assets that feed 3.2 billion individuals.

7. Changes in snow cover, lake, and stream ice, and permafrost in numerous Arctic districts have upset food supplies from crowding, hunting, fishing, and assembling exercises, hurting occupations and the social character of Arctic occupants.
8. Many practices can propel environment transformation in food frameworks, for example, disintegration control, touching area the board, hereditary enhancements for resilience to intensity and dry season, heterogeneous eating regimens, and diminished food misfortune and waste.
9. Pilot environment shrewd horticulture drives in various nations have supported efficiency, brought down outflows, further developed soil quality and water effectiveness, and expanded livelihoods and environment versatility.
10. Consumption of solid and maintainable eating regimens presents significant open doors for decreasing discharges from food frameworks and further developing wellbeing results, including through lower utilization of energy-and land-concentrated animal obtained food varieties.

6.11. HEALTH

1. Climate change is the single greatest wellbeing danger confronting humankind. The effects are as of now hurting wellbeing through air contamination, infection, outrageous climate occasions, constrained dislodging, food frailty and tensions on emotional well-being. Consistently, ecological elements end the existence of around 13 million individuals.
2. Meeting the objectives of the Paris Agreement could save around 1,000,000 lives every year overall by 2050 through decreases in air contamination alone. Staying away from the most obviously terrible environment effects could assist with forestalling 250,000 extra environment related passings each year from 2030 to 2050, fundamentally from hunger, intestinal sickness, the runs and intensity stress.
3. The worth of wellbeing gains from decreasing fossil fuel byproducts would be around twofold the worldwide expense of carrying out carbon relief measures.

4. Over 90% of individuals inhale undesirable degrees of air contamination, generally coming about because of consuming non-renewable energy sources driving environmental change. In 2018, air contamination from petroleum derivatives caused $2.9 trillion in wellbeing and financial expenses, about $8 billion every day.
5. Transportation produces around 20% of worldwide fossil fuel byproducts. Options like strolling and cycling are green as well as deal significant medical advantages, for example, lessening the risk of numerous constant ailments and working on emotional well-being.
6. Systems to deliver, bundle, and circulate food create 33% of ozone harming substance outflows. More reasonable creation would alleviate environment effects and backing more nutritious eating regimens that could forestall near 11 million unexpected losses a year.
7. Health frameworks are the principal line of guard for populaces confronted with arising wellbeing dangers, including from environmental change. To safeguard wellbeing and try not to enlarge wellbeing imbalances, nations should assemble environment versatile wellbeing frameworks.
8. The larger part of nations recognize wellbeing as really important area defenseless against environmental change. However, a tremendous money hole remains. Under 2% of multilateral environment finance goes to wellbeing projects.
9. Healthy social orders depend on well-working biological systems to give clean air, new water, drugs, and food security. These assistance to restrict sickness and settle the environment. In any case, biodiversity misfortune is going on at an extraordinary rate, influencing human wellbeing overall and expanding the risk of arising irresistible infections.

6.12. NATURE

1. Healthy biological systems can give 37% of the moderation expected to restrict worldwide temperature climb. Harmed biological systems discharge carbon as opposed to putting it away.

2. Approximately 25% of the globe's ozone harming substance emanations come from land clearing, crop creation and treatment, with animal based food contributing 75% of that.
3. With a dangerous atmospheric deviation of 1.5°C to 2°C, most of earthbound species ranges are projected to decisively shrivel. Changes in reaches can unfavorably influence species preservation, enormously increment nearby species turnover and significantly increment the risk of worldwide annihilations.
4. Climate change has been connected to more serious dangers from zoonotic sicknesses. For certain infections, expansions in temperatures or precipitation can emphatically influence the existence patterns of either the microorganism or its vector–the transitional species that spreads the sickness from the first host to people.
5. Coral reefs are especially powerless against environmental change and are projected to decline to 10–30% of previous cover at 1.5°C warming and to under 1% of previous cover at 2°C warming.
6. More than 80% of the human eating regimen is given by plants. Just three grain crops–rice, maize, and wheat–give 60% of energy admission.
7. Fish give 20% of animal protein to around 3 billion individuals.
8. Roughly 500 million individuals live in regions that experience desertification. Drylands and regions going through desertification are more defenseless against environmental change and outrageous occasions including dry spells, heatwaves, and dust storms.
9. Up to 80% of individuals living in country regions in agricultural nations depend on customary plant-based meds for essential medical services.
10. Less than 1% of complete land is utilized for mining, yet the business adversely affects biodiversity, outflows, water quality and human wellbeing.
11. The $345 billion given as worldwide appropriations to petroleum products results in $5 trillion in by and large expenses, remembering for terms of the disintegration of nature.

6.13. CAUSES OF CLIMATE CHANGE

6.13.1. Creating Power

Creating power and intensity by consuming petroleum products causes an enormous piece of worldwide emanations. Most power is as yet created by consuming coal, oil, or gas, which produces carbon dioxide and nitrous oxide–strong ozone harming substances that sweeping the Earth and trap the sun's intensity. Internationally, a smidgen in excess of a fourth of power comes from wind, sun oriented and other sustainable sources which, rather than petroleum products, transmit practically no ozone harming substances or contaminations very high. Assembling business is probably the biggest supporter of ozone depleting substance discharges around the world.

6.13.2. Chopping Down Backwoods

Chopping down backwoods to make ranches or pastures, or for different reasons, causes discharges, since trees, when they are cut, discharge the carbon they have been putting away. Every year roughly 12 million hectares of woodland are obliterated. Since timberlands ingest carbon dioxide, obliterating them likewise restricts nature's capacity to keep outflows out of the climate. Deforestation, along with farming and other land-use changes, is liable for approximately a fourth of worldwide ozone depleting substance outflows.

6.13.3. Utilizing Transportation

Most vehicles, trucks, ships, and planes run on petroleum products. That makes transportation a significant patron of ozone depleting substances, particularly carbon-dioxide emanations. Street vehicles represent the biggest part, because of the ignition of petrol based items, similar to gas, in gas powered motors. In any case, emanations from boats and planes keep on developing. Transport represents almost one fourth of worldwide energy-related carbon-dioxide emanations. What's more, patterns highlight a critical expansion in energy use for transport throughout the next few years.

6.13.4. Delivering Food

Creating food causes outflows of carbon dioxide, methane, and other ozone harming substances in different ways, including through deforestation and getting free from land for agribusiness and munching, assimilation by cows and sheep, the creation and utilization of composts and excrement for developing yields, and the utilization of energy to run ranch gear or fishing boats, for the most part with petroleum derivatives. This makes food creation a significant supporter of environmental change. Furthermore, ozone harming substance outflows likewise come from bundling and conveying food.

6.13.5. Controlling Structures

Around the world, private, and business structures consume over portion of all power. As they keep on drawing on coal, oil, and gaseous petrol for warming and cooling, they discharge huge amounts of ozone depleting substance outflows. Developing energy interest for warming and cooling, with rising forced air system proprietorship, as well as expanded power utilization for lighting, apparatuses, and associated gadgets, has added to an ascent in energy-related carbon-dioxide emanations from structures lately.

6.13.6. Consuming Excessively

Your home and utilization of force, how you move around, what you eat and the amount you discard all add to ozone depleting substance discharges. So does the utilization of merchandise like apparel, hardware, and plastics. A huge piece of worldwide ozone harming substance emanations are connected to private families. Our ways of life significantly affect our planet. The most well off bear the best liability: the most extravagant 1% of the worldwide populace consolidated represent more ozone depleting substance outflows than the least fortunate 50%.

6.14. EFFECTS OF CLIMATE CHANGE

6.14.1. Fabricating Merchandise

Assembling and industry produce emanations, for the most part from consuming petroleum products to deliver energy for making things like concrete, iron, steel, gadgets, plastics, garments, and different merchandise. Mining and other modern cycles likewise discharge gases, as does the development business. Machines utilized in the assembling system frequently run on coal, oil, or gas; and a few materials, similar to plastics, are produced using synthetic substances obtained from non-renewable energy sources.

The More sweltering temperatures

As ozone harming substance focuses rise, so does the worldwide surface temperature. The last ten years, 2011–2020, is the hottest on record. Since the 1980s, every ten years has been hotter than the past one. Essentially all land regions are seeing more hot days and intensity waves. Higher temperatures increment heat-related sicknesses and make working outside more troublesome. Out of control fires start all the more effectively and spread all the more quickly when conditions are more blazing. Temperatures in the Arctic have warmed somewhere around two times as quick as the worldwide normal.

6.14.2. More Extreme Tempests

Damaging tempests have become more extraordinary and more incessant in numerous locales. As temperatures increase, more dampness dissipates, which intensifies outrageous precipitation and flooding, causing more disastrous tempests. The recurrence and degree of typhoons is additionally impacted by the warming sea. Twisters, storms, and hurricanes feed on warm waters at the sea surface. Such tempests frequently annihilate homes and networks, causing passings and immense monetary misfortunes.

6.14.3. Expanded Dry Season

Environmental change is changing water accessibility, making it more difficult to find in additional locales. An Earth-wide temperature boost compounds water deficiencies in as of now water-pushed locales and is prompting an expanded risk of rural dry seasons influencing crops, and biological dry spells expanding the weakness of environments. Dry spells can likewise mix disastrous sand and residue storms that can get billions of lots of sand across landmasses. Deserts are extending, decreasing area for developing food. Many individuals currently face the danger of not having sufficient water consistently.

6.14.4. A Warming, Rising Sea

The sea absorbs the vast majority of the intensity from an Earth-wide temperature boost. The rate at which the sea is warming unequivocally expanded throughout recent many years, across all profundities of the sea. As the sea warms, its volume increments since water grows as it gets hotter. Dissolving ice sheets additionally cause ocean levels to rise, compromising waterfront and island networks. Moreover, the sea retains carbon dioxide, keeping it from the air. Be that as it may, more carbon dioxide makes the sea more acidic, which imperils marine life and coral reefs.

6.14.5. Loss of Species

Environmental change presents dangers to the endurance of species ashore and in the sea. These dangers increment as temperatures rise. Exacerbated by environmental change, the world is losing species at a rate multiple times more prominent than at some other time in recorded mankind's set of experiences. 1,000,000 species are in danger of becoming terminated inside the following couple of many years. Backwoods fires, outrageous climate, and obtrusive irritations and sicknesses are among numerous dangers connected with environmental change. A few animal varieties will actually want to move and make due, yet others will not.

6.14.6. Insufficient Food

Changes in the environment and expansions in outrageous climate occasions are among the explanations for a worldwide ascent in appetite and unfortunate nourishment. Fisheries, yields, and animals might be obliterated or become less useful. With the sea turning out to be more acidic, marine assets that feed billions of individuals are in danger. Changes in snow and ice cover in numerous Arctic locales have disturbed food supplies from grouping, hunting, and fishing. Heat pressure can lessen water and meadows for munching, causing declining crop yields and influencing domesticated animals.

6.14.7. More Wellbeing Risks

Environmental change is the single greatest wellbeing danger confronting mankind. Environment influences are now hurting wellbeing, through air contamination, sickness, outrageous climate occasions, constrained relocation, pressures on psychological well-being, and expanded appetite and unfortunate sustenance where individuals can't develop or track down adequate food. Consistently, natural variables end the existence of around 13 million individuals. Changing atmospheric conditions are extending infections, and outrageous climate occasions increment passings and make it hard for medical services frameworks to keep up.

6.14.8. Neediness and Dislodging

Environmental change builds the elements that put and keep individuals in neediness. Floods might clear away metropolitan ghettos, annihilating homes and jobs. Intensity can make it challenging to work in open air positions. Water shortage might influence crops. Over the course of the last ten years (2010–2019), climate related occasions uprooted an expected 23.1 million individuals on normal every year, leaving a lot more defenseless against destitution. Most evacuees come from nations that are generally powerless and least prepared to adjust to the effects of environmental change.

6.15. STATE OF THE GLOBAL CLIMATE CHANGE 2020

Groupings of the significant ozone depleting substances, CO_2, CH_4, and N_2O, kept on expanding in 2019 and 2020.

Regardless of creating La Niña conditions, worldwide mean temperature in 2020 is on course to be one of the three hottest on record. The beyond six years, including 2020, are probably going to be the six hottest years on record.

Ocean level has expanded all through the altimeter record, yet as of late ocean level has increased at a higher rate due mostly to expanded dissolving of ice sheets in Greenland and Antarctica. Worldwide mean ocean level in 2020 was like that in 2019 and both are steady with the drawn out pattern. A little drop in worldwide ocean level in the last option part of 2020 is reasonably related with creating La Niña conditions, like the impermanent drops related with past La Niña occasions.

More than 80% of the sea region experienced something like one MHW in 2020 to date. A greater amount of the sea experienced marine intensity waves 'areas of strength for named' than 'moderate' (28%).

2019 saw the most noteworthy sea heat content on record and the pace of warming throughout the last ten years was higher than the drawn out normal, demonstrating proceeded with take-up of intensity from the radiative unevenness brought about by ozone-depleting substances.

In the Arctic, the yearly least ocean ice degree was the second most reduced on endlessly record low ocean ice degrees were seen in the long periods of July and October. Antarctic ocean ice degree stayed near the drawn out normal.

The Greenland ice sheet kept on losing mass. Albeit the surface mass equilibrium was near the drawn out normal, loss of ice because of ice sheet calving was at the high finish of the 40-year satellite record. Altogether, around 152 Gt of ice were lost from the ice sheet between September 2019 and August 2020. Weighty downpour and broad flooding happened over enormous pieces of Africa and Asia in 2020. Weighty downpour and flooding impacted a large part of the Sahel, the Greater Horn of Africa, the India subcontinent and adjoining regions, China, Korea, and Japan, and portions of southeast Asia at different seasons.

With 30 named storms (starting around 17 November), the north Atlantic typhoon season had its biggest number of named storms on record, with a record number making landfall in the United States of America. The last tempest of the time (until now) Iota, was additionally the most extraordinary, arriving at classification 5.

Hurricane action in different bowls was close or beneath the drawn out mean, in spite of the fact that there were serious effects.

Serious dry season impacted many pieces of inside South America in 2020, with the most horrendously terrible impacted regions being northern Argentina, Paraguay, and western boundary areas of Brazil. Assessed rural misfortunes were close to US$3 billion in Brazil with extra misfortunes in Argentina, Uruguay, and Paraguay. Environment and climate occasions have set off huge populace developments and have seriously impacted weak individuals progressing, remembering for the Pacific district and Central America.

6.16. GLOBAL CLIMATE INDICATORS

Worldwide Climate Indicators portray the evolving environment, giving an expansive perspective on the environment at a worldwide scale. They are utilized to screen the spaces generally pertinent to environmental change, including the structure of the climate, the energy changes that emerge from the aggregation of ozone harming substances and different elements, as well as the reactions of land, sea, and ice. Subtleties on the informational collections utilized in each part can be found toward the finish of the report.

6.16.1. Temperature

The worldwide mean temperature for 2020 (January to October) was 1.2 ± 0.1°C over the 1850–1900 benchmark, utilized as a guess of pre-modern levels (Figure 6.1). 2020 is probably going to internationally be one of the three hottest years on record. The WMO evaluation depends on five worldwide temperature datasets (Figure 6.1). Every one of the five of those informational collections right now place 2020 as second hottest for the year to date when contrasted with comparable periods previously (January to October). In any case, the distinction between the main three years is little and precise rankings for every informational collection could change once the year is finished. The spread of the five appraisals for the January to October normal is between 1.11°C and 1.23°C.

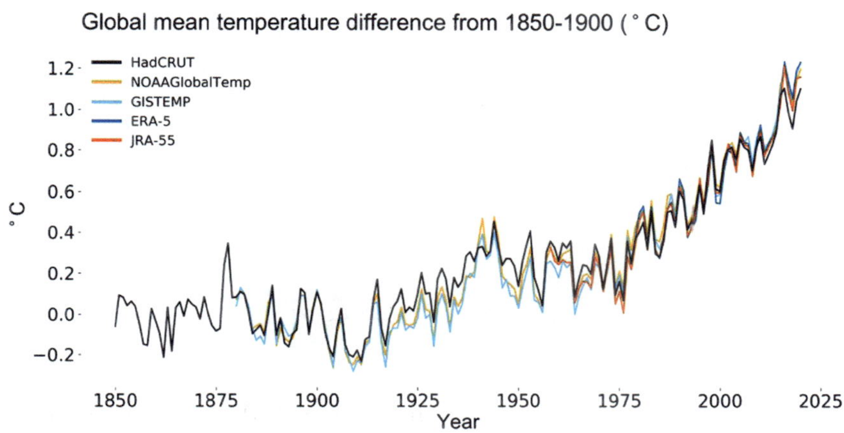

Figure 6.1: Global annual mean temperature difference from preindustrial conditions (1850–1900). The two reanalyzes (ERA5 and JRA-55) are aligned with the in situ datasets (HadCRUT, NOAAGlobalTemp, and GISTEMP) over the period 1981–2010. Data for 2020 run from January to October.

The hottest year on record to date, 2016, started with an outstandingly solid El Niño, a peculiarity which adds to raised worldwide temperatures. Notwithstanding unbiased or nearly frail El Niño conditions ahead of schedule in 20202, and La Niña conditions created by late September, the glow of 2020 is practically identical to that of 2016.

With 2020 on course to be one of the three hottest years on record, the beyond six years, 2015–2020, are probably going to be the six hottest on record. The most recent five-year (2016–2020) and 10-year (2011–2020) midpoints are likewise the hottest on record.

Albeit the general warmth of the year is clear, there were varieties in temperature abnormalities across the globe (Figure 6.2). While most land regions were hotter than the long haul (1981–2010) normal, one region in northern Eurasia stands apart with temperatures in excess of five degrees better than expected for the initial 10 months of the year (see: Sidebar: The Arctic in 2020). Other outstanding areas of warmth incorporate restricted regions of the southwestern United States, northern, and western pieces of South America, portions of Central America, and more extensive areas of Eurasia, including portions of China. For Europe, it was the hottest January to October period on record. Areas of sub optimal temperature ashore included western Canada, restricted areas of Brazil, northern India and south-eastern Australia.

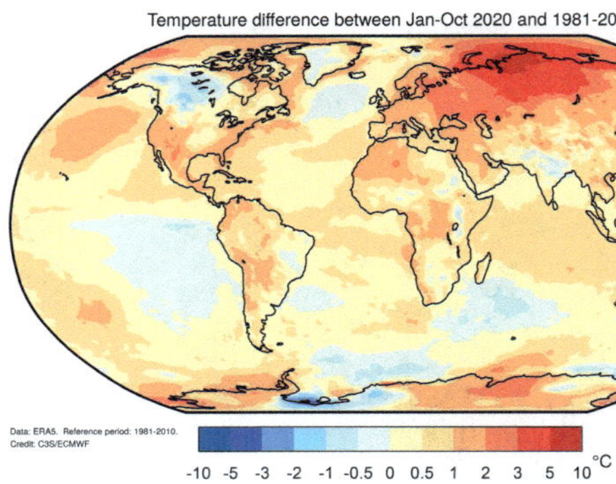

Figure 6.2: Temperature anomalies relative to the 1981–2010 long-term average from the ERA5 reanalysis for January to October 2020. Credit: Copernicus climate change service, ECMWF.

6.16.1.1. Greenhouse Gases and Stratospheric Ozone

Climatic centralizations of ozone depleting substances mirror a harmony between discharges from human exercises, regular sources, and biospheric and maritime sinks. Expanding levels of ozone depleting substances in the air because of human exercises are the significant driver of environmental change since the mid-twentieth 100 years.

Worldwide found the middle value of mole parts of ozone depleting substances are determined from in situ perceptions from various destinations in the Global Atmosphere Watch Program of WMO and accomplice organizations. These information are accessible from the World Data Center for Greenhouse Gases worked by Japan Meteorological Agency.

In 2019, ozone depleting substance focuses arrived at new highs (Figure 6.3), with all around the world found the middle value of mole parts of carbon dioxide (CO_2) at 410.5±0.2 parts per million (ppm), methane (CH_4)

at 1877±2 parts per billion (ppb) and nitrous oxide (N_2O) at 332.0±0.1 ppb. These qualities comprise, individually, 148%, 260% and 123% of pre-modern (before 1750) levels. The expansion in CO_2 from 2018 to 2019 (2.6 ppm) was bigger than both the increment from 2017 to 2018 (2.3 ppm) and the normal throughout the past 10 years (2.37 ppm each year). For CH_4, the increment from 2018 to 2019 was marginally lower than from 2017 to 2018 yet at the same time higher than the normal over the course of the past ten years. For N_2O, the increment from 2018 to 2019 was likewise lower than that saw from 2017 to 2018 and essentially equivalent to the typical development rate throughout recent years.

The transitory decrease in discharges in 2020 connected with measures taken because of COVID-195 is probably going to prompt just a slight decline in the yearly development pace of CO_2 fixation in the climate, which will be for all intents and purposes unclear from the normal interannual inconstancy driven to a great extent by the earthbound biosphere. Ongoing information from explicit areas, including Mauna Loa (Hawaii) and Cape Grim (Tasmania) show that degrees of CO_2, CH_4 and N_2O kept on expanding in 2020.

The IPCC Special Report on Global Warming of 1.5°C found that restricting warming to 1.5°C above pre-modern levels infers arriving at net-zero CO_2 discharges worldwide around 2050 and simultaneous profound decreases in emanations of non-CO_2 forcers.

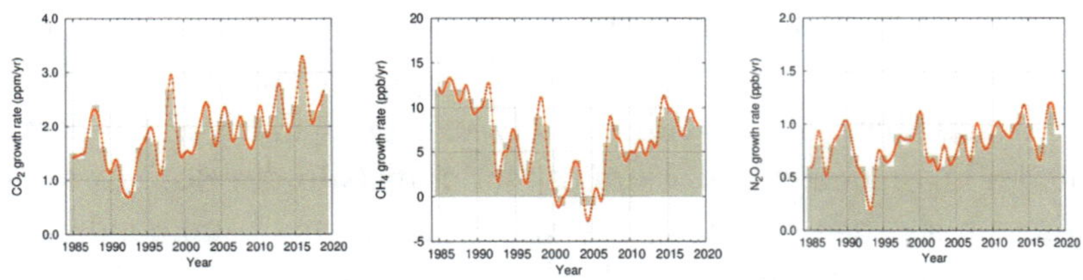

Figure 6.3: Top row: Globally averaged mole fraction (measure of concentration), from 1984 to 2019, of CO_2 in parts per million (left), CH_4 in parts per billion (center) and N_2O in parts per billion (right). The red line is the monthly mean mole fraction with the seasonal variations removed; the blue dots and line show the monthly averages. Bottom row: the growth rates representing

increases in successive annual means of mole fractions for CO_2 in parts per million per year are shown as gray columns (left), CH_4 in parts per billion per year (center) and N_2O in parts per billion per year (right) (Source: WMO Global Atmosphere Watch).

6.16.1.2. Stratospheric Ozone and Ozone-Depleting Gases

Following the outcome of the Montreal Protocol, utilization of halons and CFCs has been accounted for as ended however their levels in the environment keep on being observed. On account of their long lifetime, these mixtures will stay in the air for a long time and regardless of whether there were no new discharges, there is even a very sizable amount of chlorine and bromine present to cause total annihilation of ozone in Antarctica from August to December. Accordingly, the arrangement of the Antarctic ozone opening keeps on being a yearly spring occasion with year-to-year variety in its size and profundity administered generally by meteorological circumstances.

The 2020 ozone opening grew somewhat early and kept developing, bringing about a huge and profound ozone opening which, as of November 2020, has a bigger than normal (1979–2019 mean) area of ozone exhaustion and below the norm least ozone. The ozone opening region arrived at its greatest region for 2020 on 20 September at 24.8 million km^2, a similar region as in 2018. The region of the opening was nearer to the maxima seen in years, for example, 2015 (28.2 million km^2) and 2006 (29.6 million km^2) than 2019 (16.4 million km^2) as indicated by an examination from the National Aeronautics and Space Administration (NASA).

At the opposite finish of the earth, surprising air conditions prompted ozone focuses over the Arctic tumbling to a record low for the long stretch of March. Surprisingly powerless "wave" occasions in the upper air left the polar vortex generally undisturbed, keeping blending of ozone-rich air from lower scopes. What's more, the stratospheric polar vortex over the Arctic was solid and this, joined with reliably exceptionally low temperatures, permitted a huge area of polar stratospheric mists to develop. At the point when the sun ascends after the polar winter, it triggers compound cycles in the polar stratospheric mists that lead to exhaustion of ozone. Estimations from weather conditions inflatables showed ozone exhaustion outperforming the levels revealed in 2011 and along with satellite perceptions highlight stratospheric ozone levels down to around 205 Dobson Units on March 12, 2020. The regular most reduced ozone esteems recently saw over the Arctic in March are somewhere around 240 Dobson Units.

6.16.2. Ocean

Most of the overabundance energy that amasses in the earth framework because of expanding centralizations of ozone depleting substances is taken up by the sea. The additional energy warms the sea and the resulting warm development of the water prompts ocean level ascent. The surface sea has warmed more quickly than the inside and this should be visible in the ascent of worldwide mean temperature and furthermore in the expanded occurrence of MHWs. As the centralization of CO_2 in the air rises, so too does the grouping of CO_2 in the seas. This influences sea science, bringing down the normal pH of the water, a cycle known as sea fermentation. This large number of changes have wide scope of effects in the seas and beach front regions.

6.16.2.1. Ocean Heat Content

Expanding human outflows of CO_2 and other ozone harming substances cause a positive radiative irregularity at the highest point of the environment–the Earth Energy Imbalance (EEI)–which is driving a worldwide temperature alteration through an aggregation of energy as intensity in the Earth system Sea Heat Content (OHC) is a proportion of this intensity gathering in the Earth framework, as around 90% of it is put away in the sea. A positive EEI signals that the Earth's environment framework is as yet answering the current forcing and that really warming will happen regardless of whether the compelling increment further.

Verifiable estimations of subsurface temperature back to the 1940s generally depend on shipboard estimation frameworks, which compels the accessibility of subsurface temperature perceptions at a worldwide scale and at depth. With the organization of the Argo organization of independent profiling floats, which previously accomplished close worldwide inclusion in 2006, it is currently conceivable to regularly gauge OHC changes down to a profundity of 2000m.

Different examination bunches have created evaluations of worldwide OHC. In spite of the fact that they all depend pretty much on similar data set, the evaluations show contrasts emerging from the different measurable medicines of information holes, the decision of climatology and the methodology used to represent instrumental biases. A coordinated exertion has been laid out to give a worldwide view on the worldwide advancement of sea warming, and an update to 2019 is displayed in Figures 6.4 and 6.5. More subtleties can be found in von Schuckmann et al. (2020).

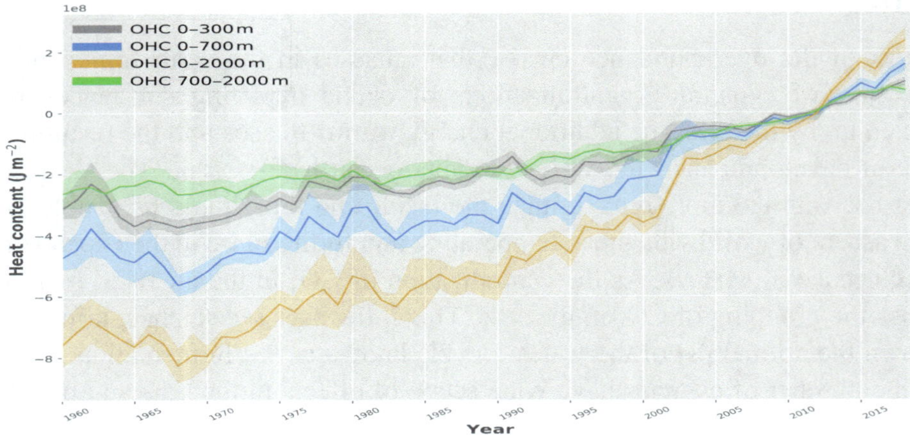

Figure 6.4: 1960–2019 ensemble meantime series and ensemble standard deviation (2-sigma, shaded) of global ocean heat content (OHC) anomalies relative to the 2005–2017 climatology for the 0–300 m (gray), 0–700 m (blue), 0–2000 m (yellow) and 700–2000 m depth layer (green). The ensemble mean is an outcome of a concerted international effort, and all products used are listed at the end of the report and in the legend of Figure 6.5. Note that values are given for the ocean surface area between 60°S-60°N, and limited to the 300m bathymetry of each product. Updated from von Schuckmann et al. (2020).

The 0–2000m profundity layer of the worldwide sea kept on warming in 2019 arriving at another record high (Figure 6.4), and it is normal that it will keep on warming in the future. Heat capacity at halfway profundity (700–2000 m) expanded at a practically identical rate to the 0–300 m profundity layer, which is overall understanding among the 15 worldwide OHC gauges (Figure 6.5). All informational collections concur that sea warming rates show an especially amazing expansion in the beyond twenty years. Also, there is an obvious sign that heat sequestration into the sea beneath 700m profundity occurred throughout the course of recent many years connected to an increment of OHC patterns over the long haul. Sea warming rates for the 0–2000 m profundity layer arrived at record paces of 1.2 (0.8) ± 0.2 Wm^{-2} for the sea (worldwide) region over the period 2010–2019. Beneath 2000 m profundity, the sea additionally warmed but at the lower pace of 0.07 ± 0.04 Wm^{-2} from 1991–2018.

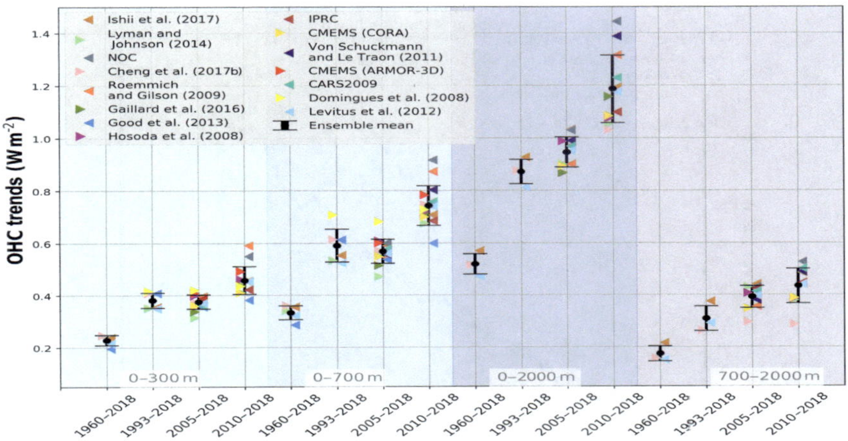

Figure 6.5: Linear trends of global ocean heat content (OHC) as derived from different temperature products (colors). References are provided at the end of the report. The ensemble mean and standard deviation (2-sigma) is given in black, respectively. The shaded areas show trends from different depth layer integrations, i.e. 0–300 m (light turquoise), 0–700 m (light blue), 0–2000 m (purple) and 700–2000 m (light purple). For each integration depth layer, trends are evaluated over four periods: historical (1960–2019), altimeter era (1993–2019), golden Argo era (2005–2019), and the most recent period 2010–2019. Updated from von Schuckmann et al. (2020).

6.16.2.2. Sea Level

By and large, since mid-1993, the altimetry-based worldwide mean pace of ocean level ascent adds up to 3.3 ± 0.3 mm/yr. The rate has additionally expanded throughout that time. A more noteworthy loss of ice mass from the ice sheets is the primary driver of the sped-up ascent in the worldwide mean ocean level.

The worldwide mean ocean level in 2020 has been like that in 2019 (Figure 6.6, left). A little reduction in the last option part of 2020 is reasonably connected with La Niña conditions in the tropical Pacific. Interannual changes of worldwide mean ocean level around the drawn out pattern are corresponded with ENSO fluctuation (Figure 6.6, right). During La Niña occasions, for example, that in late 2020 and the solid La Niña of 2011, changes in precipitation designs move water mass from the sea to tropical stream bowls ashore, briefly lessening worldwide mean ocean level. The inverse is seen during El Niño (for example, the solid 2015–16 El Niño).

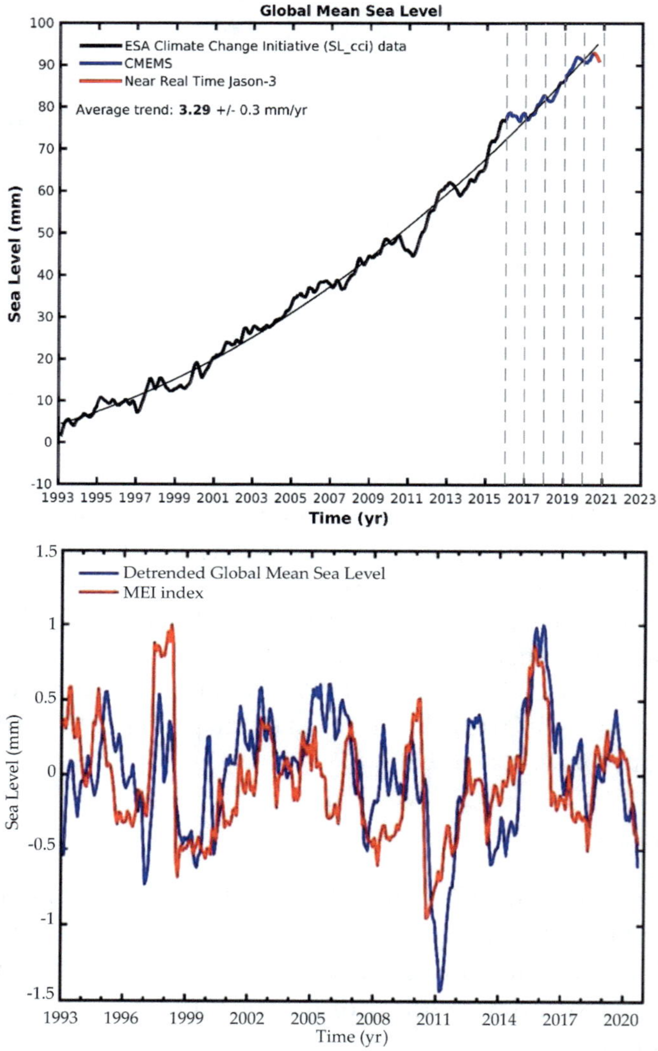

Figure 6.6: (left) Satellite altimetry-based global mean sea level for January 1993 to October 2020 (last data: 13 October 2020). Data from the ESA climate change Initiative sea-level project, from January 1993 to December 2015, (thick black curve); extended by the Copernicus Marine and Environment Service, CMEMS, until August 2020 (blue curve) and with near real time altimetry data from the Jason-3 mission beyond August 2020 (red curve). The thin black curve is a quadratic function that best fits the data. Vertical dashed lines mark the start of each year from 2016 to 2021. (right) Interannual variability of the (with the quadratic function shown in left-hand panel subtracted) global mean sea level (blue curve) with the multivariate ENSO index (red curve) superimposed.

At a provincial scale, ocean level keeps on rising non-consistently. The most grounded local patterns over the period January 1993 to January 2020 are found in the southern half of the globe: east of Madagascar in the Indian Ocean; east of New Zealand in the Pacific Ocean; and east of Rio de la Plata/ South America in the south Atlantic. A lengthened toward the east example is additionally found in the North Pacific. A solid example that was found in the western tropical Pacific over the initial twenty years of the altimetry record is currently blurring, recommending that it was connected with momentary changeability. Territorial ocean level patterns are overwhelmed by varieties in sea heat content. Be that as it may, in certain locales, for example, the Arctic, saltiness changes because of freshwater input from the dissolving of ice ashore assume a significant part.

6.16.2.3. Marine Heatwaves

As with heatwaves ashore, outrageous intensity can influence the close surface layer of the seas with a scope of ramifications for marine life and ward networks. Satellite recoveries of ocean surface temperature can be utilized to screen MHWs. MHWs are arranged here as Moderate, Strong, Severe or Extreme (for definitions see MHW information).

A significant part of the sea experienced no less than one 'In number' MHW sooner or later in 2020 (Figure 6.7a). Prominently missing are MHWs in the Atlantic Ocean south of Greenland, and in the eastern central Pacific. The Laptev Sea encountered an 'Outrageous' MHW from June to October. Ocean ice degree was strangely low in the locale and neighboring area regions experienced heatwaves throughout the late spring. Another significant MHW to note in 2020 was the arrival of the semi-tireless warm district in the upper east Pacific. This occasion is comparable in scale to the first blob which created around 2013 with leftovers going on until 201623. Roughly one fifth of the worldwide sea was encountering a MHW on some random day in 2020 (Figure 6.7b). This is like 2019, however not exactly the 2016 pinnacle of 23%. A greater amount of the sea experienced MHWs named 'solid' (43%) than 'moderate' (28%). Altogether, 82% of the sea experienced somewhere around one MHW during 2020 (until now, Figure 6.7c), which is not exactly both 2019 (84%), and the 2016 pinnacle (88%).

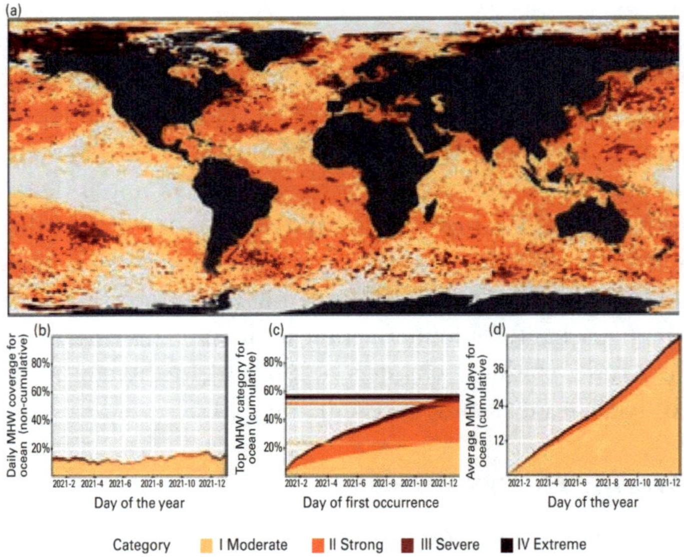

Figure 6.7: (A) Global map showing the highest MHW category (for definitions see Marine heatwave data) experienced at each pixel over the course of the year, estimated using the NOAA OISST v2.1 dataset (reference period 1982–2011). Light gray indicates that no MHWs occurred in a pixel over the entire year; (B) Stacked bar plot showing the percentage of ocean pixels experiencing an MHW on any given day of the year; (C) Stacked bar plot showing the cumulative percentage of the ocean that experienced an MHW over the year.. Horizontal lines in this figure show the final percentages for each category of MHW; (D) Stacked bar plot showing the cumulative number of MHW days averaged over all pixels in the ocean [25](Source: Robert Schlegel, IMEV).

6.16.2.4. Ocean Acidification

The sea assimilates around 23% of the yearly discharges of anthropogenic CO_2 to the atmosphere, accordingly assisting with reducing the effects of environmental change on the planet. The biological expenses of this interaction to the sea are high, as the CO_2 responds with seawater, bringing down its pH, a cycle known as sea fermentation. Sea fermentation influences numerous animals and biological system administrations, compromising food security by imperiling fisheries and hydroponics. This is especially an issue in the polar seas as a result of the sea science of these chilly districts. It additionally influences beach front insurance by debilitating coral reefs, which safeguard shorelines. As the acridity of the sea expands, its ability to ingest CO_2 from the air diminishes, hampering the sea's job in directing environmental change. Ordinary worldwide perception and estimation of sea fermentation is expected to work on the comprehension of its ramifications, empower demonstrating and forecast of progress and inconstancy, and assist with illuminating alleviation and variation procedures.

Worldwide endeavors have been made to gather and analyze sea fermentation perception information, which contribute towards the Sustainable Development Goal (SDG) 14.3 and the related SDG Indicator 14.3.1: "Normal marine causticity (pH) estimated at concurred set-up of agent examining stations.

The information is summed up in Figure 6.8 (left) and show an increment of changeability (least and greatest pH values are featured) and a decrease in normal pH at the accessible noticing destinations somewhere in the range of 2015 and 2019. The more consistent worldwide change (Figure 6.8, right) assessed from a more extensive assortment of sources including estimations of different factors, diverges from the local and occasional varieties in sea carbonate science seen at individual destinations. The rising number of accessible information features the changeability and the pattern in sea fermentation and the requirement for supported long haul perceptions to more readily describe the normal inconstancy in sea carbonate science. Long-term observations to better characterize the natural variability in ocean carbonate chemistry.

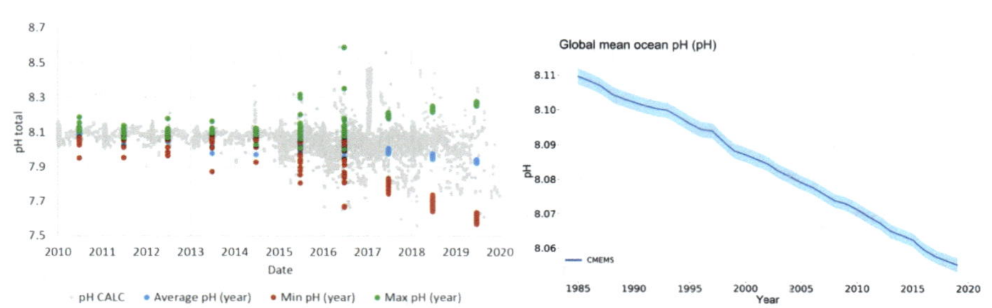

Figure 6.8: (left) Surface pH values based on ocean acidification data submitted to the 14.3.1 data portal (http://oa.iode.org) for the time period from 1 January 2010 to 8 January 2020. Grey circles–calculated pH of data submissions (including all data sets with data for at least two carbonate parameters); blue circles–average annual pH (based on data sets with data for at least two carbonate parameters); red circles–annual minimum pH; green circles–annual maximum pH. Note that the number of stations is not constant with time. (right) Global mean surface pH from E.U. Copernicus Marine Service Information (blue). The shaded area indicates the estimated uncertainty in each estimate.

6.16.3. Cryosphere

The cryosphere is the area that contains the frozen pieces of the earth. The cryosphere gives key marks of the evolving environment, however it is one of the most under-inspected spaces. The major cryosphere pointers utilized in the temporary proclamation are ocean ice degree and mass equilibrium of the Greenland ice sheets, the two of which can be regularly estimated utilizing satellite information. Explicit snow occasions are canvassed in the segment on High-influence occasions in 2020. Other cryosphere markers including ice sheet mass equilibrium and permafrost, which are estimated in situ, are not shrouded in the temporary proclamation as it requires investment to assemble and deal with the information.

6.16.3.1. Sea Ice

In the Arctic, the yearly least ocean ice degree in September was the second most reduced on record (Figure 6.9) and record low ocean ice degrees were seen in the long stretches of July and October. April and August degrees were among the five least in the 42-year satellite information record. Antarctic ocean ice stayed near the drawn out normal. For additional subtleties on the datasets utilized see Sea ice information.

In the Arctic, the greatest ocean ice degree for the year was reached on 5 March 2020. At simply over 15 million km^2, this was the tenth or eleventh (contingent upon the informational collection utilized) least greatest degree on record. Ocean ice retreat in late March was for the most part in the Bering Sea. In April, the pace of decline was like that of late years, and the mean ocean ice degree for April was somewhere in the range of second and fourth most minimal, successfully attached with 2016, 2017, and 2018.

Record high temperatures north of the Arctic Circle in Siberia (see Sidebar: The Arctic in 2020) set off a speed increase of ocean ice liquefy in the East Siberian and Laptev Seas, which proceeded with well into July. Ocean ice degree for July was the most minimal on record (7.28 million km^2). The ocean ice retreat in the Laptev Sea was the earliest seen in the satellite period. Towards the finish of July, a typhoon entered the Beaufort Sea and spread the ocean ice out, briefly easing back the abatement of the ice degree. In mid-August, the region impacted by the tornado softened quickly which, joined with the supported liquefy in the East Siberian and Laptev Seas, made the August degree the second or third least on record.

The 2020 Arctic ocean ice degree least was seen on 15 September at 3.74 million km^2, stamping just the subsequent time on record that the Arctic ocean ice degree shrank to under 4 million km^2. Just 2012 had a lower least degree at 3.39 million km^2. Huge areas of untamed sea were seen in the Chukchi, East Siberian, Laptev, and Beaufort Seas, despite a tongue of long term ice that endure the 2020 soften season in the Beaufort Sea Figure 6.10). Refreeze was delayed in late September and October in the Laptev and East Siberian Seas, most likely because of the aggregated intensity in the upper sea since the early retreat in late June. Cold ocean ice degree was the least on record for October. In the Antarctic, the base ocean ice degree was noticed later than expected, on 2 March, at 2.73 million km^2. Ocean ice degree was near the climatological normal for the vast majority of the year and was better than expected in September and October. The most extreme degree was reached on 28 September at 18.95 million km^2.

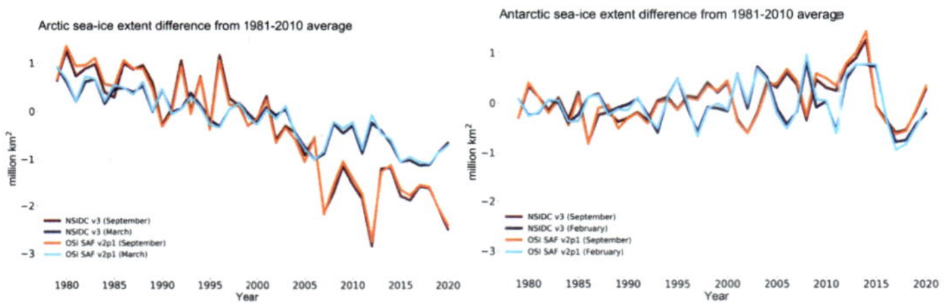

Figure 6.9: Sea-ice extent difference from the 1981–2010 average in the Arctic (left) and Antarctic (right) for the months with maximum ice cover (Arctic: March, Antarctic: September) and minimum ice cover (Arctic: September, Antarctic: February). Data from EUMETSAT OSI SAF v2p1 (Lavergne et al., 2019) and NSIDC v3 (Fetterer et al., 2017).

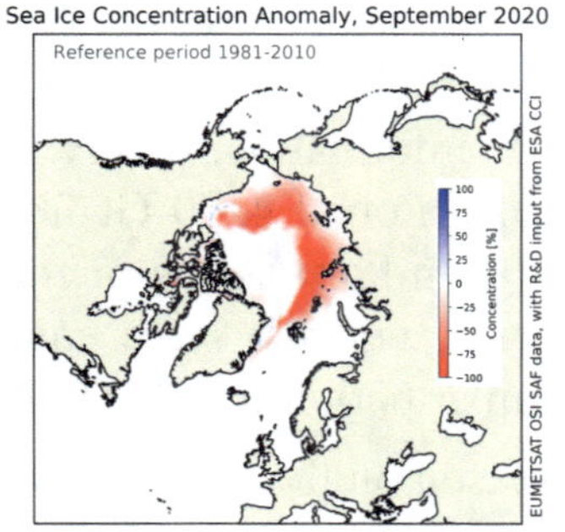

Figure 6.10: Sea-ice concentration anomaly against the 1981–2010 period for the Arctic (EUMETSAT OSI SAF v2p1 data, with R&D input from ESA CCI). Sea-ice concentration is a parameter used to describe how densely the sea ice is packed. It is expressed by a continuous scale from complete ice cover (100%) to an open ocean with no ice (0%).

6.16.3.2. Greenland Ice Sheet

Notwithstanding the excellent warmth in enormous pieces of the Arctic, specifically the extremely surprising temperatures that were seen in eastern Siberia since February 2020, temperatures over Greenland were near the drawn out mean (Figure 6.2). The ice sheet finished the September 2019 to August 2020 season with a general deficiency of 152 Gt (gigatons) of ice from surface dissolving, the release of ice shelves and from softening of icy mass tongues by warm seawater (Figure 6.11). This implies that the ice sheet kept on losing ice, however at a more slow rate than seen in 2019 (which saw a deficiency of 329 Gt).

Changes in the mass of the Greenland ice sheet mirror the consolidated impacts of surface mass equilibrium (SMB)–characterized as the distinction among snowfall and run-off from the ice sheet, which is generally certain toward the year's end–and mass misfortunes at the outskirts from the calving of ice shelves and the liquefying of glacial mass tongues that meet the sea. The 2019–20 Greenland SMB was +349 Gt of ice, which is near the 40-year normal of +341 Gt. Be that as it may, ice misfortune because of icy mass calving was at the high finish of the 40-year satellite record. The Greenland SMB record is presently forty years in length and, in spite of the fact that it fluctuates over time one year to another, there has been a general decrease in the normal SMB over the long haul (Figure 6.11). During the 1980s and 1990s, the typical SMB gain was about +416 Gt/year. It tumbled to +270 Gt during the 2000s and +260 Gt during the 2010s.

The GRACE satellites and the subsequent mission GRACE-FO measure the small difference in the gravitational power because of changes in how much ice. This provides us with a free proportion of the absolute mass equilibrium. In view of this information, it very well may be seen that the Greenland ice sheet has lost around 4200 Gt from April 2002 to August 2019, which added to an ocean level ascent of somewhat more than 1 cm. This is in great concurrence with the mass equilibrium from SMB and release, which during a similar period was 4261 Gt.

The 2019–20 soften season on the ice sheet began on 22 June, 10 days after the fact than the 1981–2020 normal. As in earlier years, there were misfortunes along the Greenlandic West coast and gains in the east. In mid-August, bizarrely huge tempests brought multiple times the typical month to month precipitation to western Greenland, the majority of which fell as snow that briefly halted the total deficit of ice and was conclusive in lessening how much soften, very not the same as the earlier year 2018–19 with broadened high strain periods and a lot of daylight which altogether expanded how much liquefy in summer.

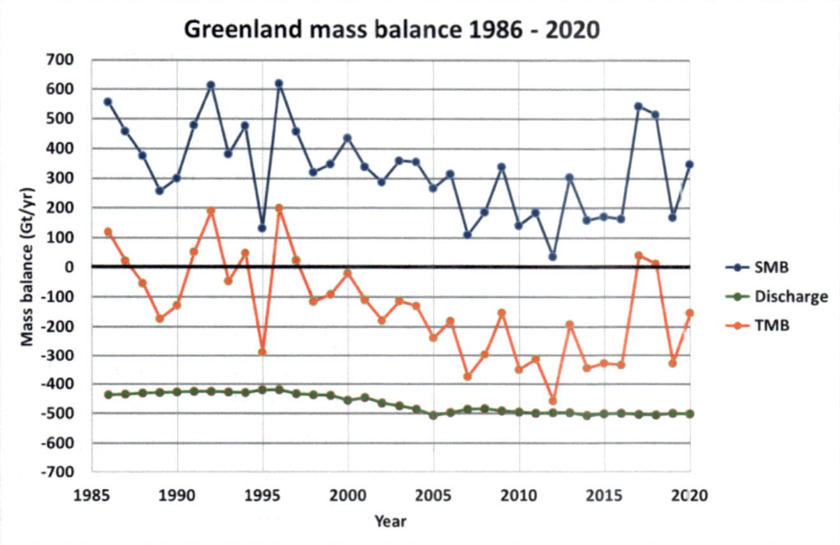

Figure 6.11: Components of the total mass balance of the Greenland Ice Sheet 1986–2020. Blue: Surface mass balance, green: discharge, red: total mass balance, the sum of Surface Mass Balance and discharge.

6.16.4. Drivers of Short-Term Climate Variability

There are various normal peculiarities, frequently alluded to as environment examples or environment modes, that influence climate at timescales going from days to a while. Surface temperatures change generally leisurely over the sea, so repeating designs in ocean surface temperature can be utilized to comprehend and, sometimes, anticipate the more quickly changing examples of weather conditions throughout land on occasional time scales. Additionally, though at a quicker rate, realized pressure changes in the climate can assist with making sense of specific territorial atmospheric conditions.

In 2020, the El Niño-Southern Oscillation, a peculiarity estimated by both the seas and the environment, and the Arctic Oscillation (AO), a climatic peculiarity, each added to explicit climate occasions in various areas of the planet. The Indian Ocean Dipole, which assumed a vital part in the occasions of 2019 was close unbiased for quite a bit of 2020.

6.16.4.1. El Niño Southern Oscillation (ENSO)

ENSO is one of the main drivers of year-to-year changeability in weather conditions all over the planet. It is connected to perils like weighty downpours, floods, and dry spell. El Niño, described by higher-than-normal ocean surface temperatures in the eastern Pacific and a debilitating of the exchange twists, regularly impacts worldwide temperatures. La Niña, which is portrayed by less than ideal ocean surface temperatures in the focal and eastern Pacific and a fortifying of the exchange winds, makes the contrary difference.

Ocean surface temperatures toward the finish of 2019 were near or surpassed El Niño edges in the Niño 3.4 district. These continued into the early long periods of 2020, yet the occasion didn't fortify and ocean surface temperature abnormalities in the eastern Pacific fell in March. Following a six-month time of nonpartisan circumstances–that is, ocean surface temperatures inside 0.5°C of typical–the cool-stage, La Niña, created in August and reinforced in October, despite the fact that it was actually still characterized as frail (0.5–1.0°C underneath ordinary). The air likewise answered with more grounded than normal exchange winds, showing a coupling with the ocean surface temperatures. La Niña conditions are related with better than expected typhoon movement in the North Atlantic, which has encountered a record number of named hurricanes during its 2020 tropical storm season. La Niña is supposed to continue through the primary quarter of 2021.

6.16.4.2. Arctic Oscillation (AO)

The AO is a huge scope barometrical example that impacts climate all through the Northern Hemisphere. The positive stage is portrayed by below the norm gaseous tension over the Arctic and higher-than-normal strain over the northern Pacific and Atlantic Oceans. The fly stream is lined up with the lines of scope and farther north than normal, securing cold Arctic air, and tempests can be moved toward the north of their standard ways. The mid-scopes of North America, Europe, Siberia, and East Asia for the most part see less virus air episodes than expected during the positive period of the AO. A negative AO makes the contrary difference, related with a seriously wandering plane stream and cold air spilling south into the mid-scopes.

The AO was firmly sure during the Northern Hemisphere 2019–20 winter and was the most grounded in February since January 1993. This added to the hottest winter on record for Asia and Europe and the 6th hottest for the adjoining United States, differentiated by Alaska's coldest winter in over twenty years. Furthermore, the positive winter period of the AO has been connected to low ocean ice degree the accompanying summer (Rigor et al., 2002). See the segment on Sea ice.

6.16.4.3. Sidebar: The Arctic in 2020

The Arctic is going through extraordinary changes as the worldwide temperature increments. Since the mid-1980s, Arctic surface air temperatures have warmed no less than two times as quick as the worldwide normal, while ocean ice, the Greenland ice sheet, and icy masses have declined over a similar period and permafrost temperatures have expanded. This has possibly enormous ramifications for Arctic environments, yet additionally for the worldwide environment through different feedbacks.

For the initial 10 months of 2020, the Arctic stands apart as the area with the biggest temperature deviations from the drawn out normal. Differentiating states of ice, intensity, and out of control fires were found in the eastern and western Arctic (Figure 6.12). An unequivocally certain period of the AO during Winter 2019–20 set everything up right off the bat in the year, with higher than normal temperatures across Europe and Asia and well-less than ideal temperatures in Alaska, an example which endured through a large part of the year.

In a huge locale of the Siberian Arctic, temperature peculiarities for January to October were multiple degree centigrade, and in its focal waterfront parts in excess of 5°C, better than expected (Figure 6.14). A fundamental record temperature was set for north of the Arctic Circle, of 38°C on 20 June in Verkhoyansk during a drawn out heatwave. Heatwaves and heat records were additionally seen in different pieces of the Arctic (see High-influence occasions in 2020) and outrageous intensity was not restricted to the land as it were.

An "outrageous" MHW impacted huge region of the Arctic Ocean north of Eurasia (Figure 6.7). Ocean ice in the Laptev Sea, seaward from the area of most elevated temperature peculiarities ashore, was abnormally low through the mid-year and harvest time. To be sure, ocean ice degree was especially low along the Siberian shoreline, with the Northern Sea Route sans ice or near ice liberated from July to October.

Albeit the Arctic was transcendently hotter than normal for this period, a few areas, including portions of Alaska and Greenland, saw near or less than ideal temperatures. Subsequently, the 2019–20 surface mass equilibrium for Greenland was near the 40-year normal. By and by, the decay of the Greenland ice sheet went on during the 2019–20 season, yet the misfortune was underneath the normal sums seen during the last 10 years (see Cryosphere). Ocean ice conditions along the Canadian archipelago were near normal at the September least and the western section remained closed.

The fierce blaze season in the Arctic during 2020 has been especially dynamic, however with enormous provincial contrasts. The district north of the Arctic circle saw the most dynamic fierce blaze season in a 18-year information record, as assessed as far as fire radiative power and CO_2 discharges set free from flames. The fundamental action was moved in the eastern Siberian Arctic, which was additionally drier than normal.

Provincial reports for eastern Siberia demonstrate that the timberland fire season began sooner than normal, and for certain areas finished later, bringing about long haul harm to neighborhood environments. The Frozen North, as well as Yukon and the Northwest Territories detailed fire movement that was well sub optimal.

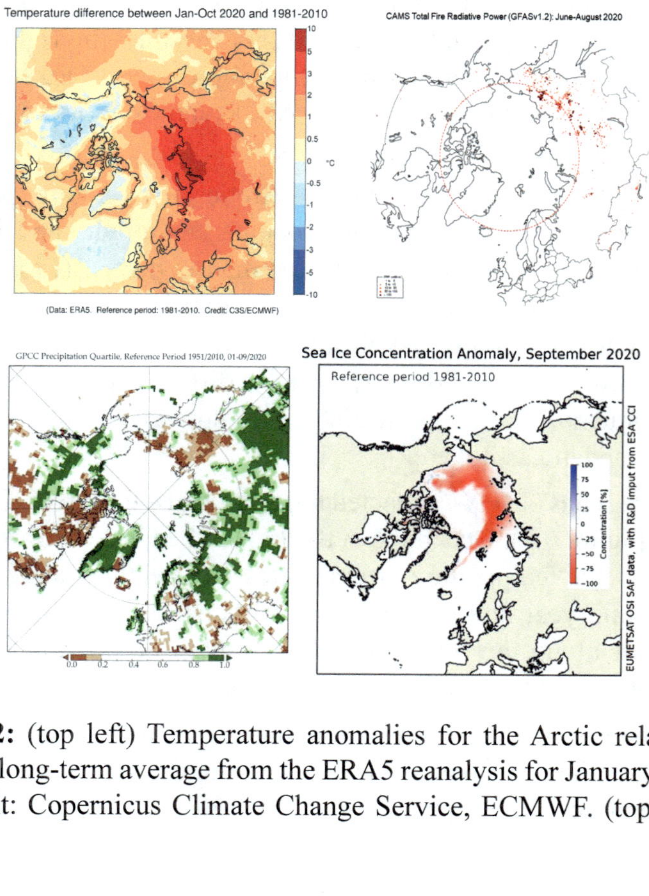

Figure 6.12: (top left) Temperature anomalies for the Arctic relative to the 1981–2010 long-term average from the ERA5 reanalysis for January to October 2020. Credit: Copernicus Climate Change Service, ECMWF. (top right) Fire

Radiative Power, a measure of heat output from wildfires, in the Arctic Circle between June and August 2020. Credit: Copernicus Atmosphere Monitoring Service, ECMWF (bottom left) Total precipitation in Jan-Sep 2020, expressed as a percentile of the 1951–2010 reference period, for areas that would have been in the driest 20% (brown) and wettest 20% (green) of years during the reference period, with darker shades of brown and green indicating the driest and wettest 10%, respectively Credit: Global Precipitation Climatology Centre (GPCC). (bottom right) sea-ice concentration anomaly for September 2020. Credit: EUMETSAT OSI SAF data with R&D input from ESA CCI.

6.17. HIGH-IMPACT EVENTS IN 2020

Albeit understanding wide scale changes in the environment is significant, the most intense effects of climate and environment are many times felt during outrageous meteorological occasions like weighty downpour and snow, dry spells, heatwaves, cold waves, and tempests, including typhoons. These can prompt or compound other high-influence occasions like flooding, avalanches, rapidly spreading fires and torrential slides.

6.17.1. A Year of Widespread Flooding, Especially in Africa and Asia

Extremely broad flooding happened over huge pieces of Africa in 2020. Precipitation was well better than expected in the vast majority of the Greater Horn of Africa during the March-May "long rains" season, following a correspondingly wet season in October-December 2019. This was trailed by better than expected precipitation across by far most of the Sahel district, from Senegal to Sudan, throughout the late spring storm (Figure 6.13.).

Flooding was broad across many pieces of the district, in spite of the fact that Sudan and Kenya were the most awful impacted with 285 passings revealed in Kenya, and 155 passings and more than 800,000 individuals impacted in Sudan.

Nations revealing death toll or huge uprooting of populaces included Sudan, South Sudan, Ethiopia, Somalia, Kenya, Uganda, Chad, Nigeria, Niger, Benin, Togo, Senegal, Côte d'Ivoire, Cameroon, and Burkina Faso. Numerous lakes and waterways arrived at record undeniable levels, remembering Lake Victoria for May, and the Niger River at Niamey and the Blue Nile at Khartoum in September.

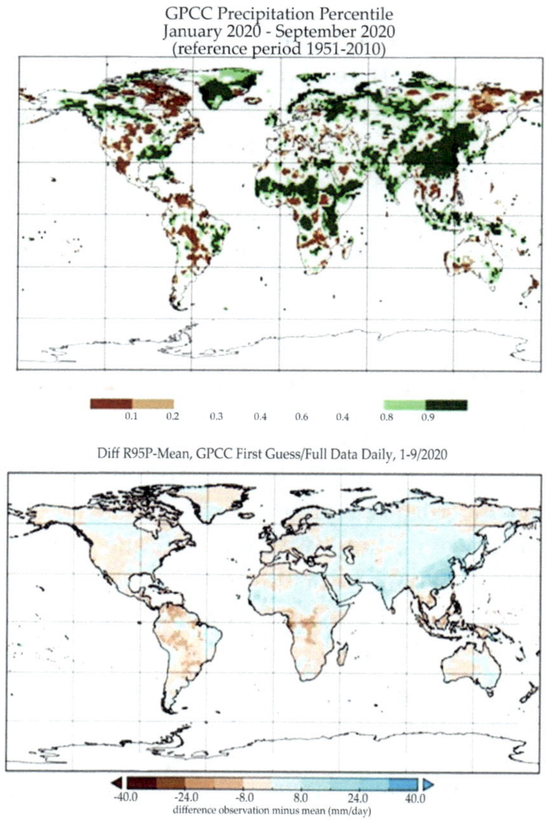

Figure 6.13: (left) Total precipitation in Jan-Sep 2020, expressed as a percentile of the 1951–2010 reference period, for areas that would have been in the driest 20% (brown) and wettest 20% (green) of years during the reference period, with darker shades of brown and green indicating the driest and wettest 10%, respectively (Source: Global Precipitation Climatology Centre (GPCC), Deutscher Wetterdienst, Germany). (right) Difference between the observed 95th percentile of daily precipitation total in Jan-Sep 2020 and the long-term mean based on 1982–2016 (full year). Blue indicate more extreme daily precipitation events and brown less than the long-term means.

India had one of its two wettest storm seasons starting around 1994, with broadly found the middle value of precipitation for June to September 9% over the drawn out normal. Weighty downpour, flooding, and avalanches additionally impacted encompassing nations. August was the wettest month on record for Pakistan, and 231 mm of downpour fell on 28 August at Karachi-Faisal, the most elevated everyday absolute on record in the Karachi region. In excess of 2000 passings were accounted for during the season in India, Pakistan, Nepal, Bangladesh, Afghanistan, and Myanmar, remembering 145 passings for streak flooding in Afghanistan in late August, and 166 passings in an avalanche at a mine in Myanmar toward the beginning of July, following weighty downpour.

Relentless high precipitation in the Yangtze River catchment in China in the rainstorm season likewise caused extreme flooding. The June-July period was especially wet and floods impacted the Yangtze and its feeders, with the Three Gorges Dam releasing water at its greatest limit. Revealed financial misfortunes surpassed US$15 billion, and something like 279 passings were accounted for during the period. It was additionally an exceptionally wet summer rainstorm season over the Korean Peninsula, with the Republic of Korea encountering its third-wettest summer, and portions of western Japan were impacted by critical flooding in July.

Portions of southeast Asia experienced serious flooding in October and November. The most terrible impacted region was focal Vietnam, where weighty downpours common of the appearance of the upper east rainstorm were exacerbated by a progression of hurricanes and sorrows, with eight making landfall in under five weeks. Hué got more than 1800 mm of downpour in the week from 7 to 13 October, and a sum of 2615 mm for the long stretch of October. The flooding additionally broadened further west into Cambodia.

Different districts to encounter critical death toll in floods (for the most part streak floods related with outrageous neighborhood precipitation) remembered Indonesia for January, Brazil in January and March, the Democratic Republic of the Congo and Rwanda in April and May, and Yemen in July. Jakarta had its wettest day starting around 1996 with 377 mm at Halim Airport on 1 January, and Belo Horizonte its wettest day on record with 172 mm on 24 January. Khombole in Senegal had its wettest day on record with 225.8 mm on 5 September.

6.17.2. Heatwaves, Drought, and Wildfire

Extreme dry spell impacted many pieces of inside South America in 2020, with the most obviously awful impacted regions being northern Argentina, Paraguay, and western boundary areas of Brazil. It was the second-driest January-August period on record for the upper east district of Argentina, while precipitation for the period was likewise far sub optimal in Paraguay. Assessed agrarian misfortunes were close to US$3 billion in Brazil with extra misfortunes in Argentina, Uruguay, and Paraguay. Peru additionally experienced dry season conditions among January and March, for the most part in the north of the country. As the dry spell proceeded, a significant heatwave extended across the locale in late September and early October, stretching out east and north to cover quite a bit of inside Brazil. It came to 44.6°C on 5 October at Nova Maringá and Água Clara, while individual areas which had their most sultry day on record included Cuiaba, Curitiba, Belo Horizonte and Asuncion. There was critical fierce blaze action across each of the three nations from mid-year onwards, with probably the main out of control fires happening in the Pantanal wetlands in western Brazil.

2020 has been an extraordinarily warm year in the vast majority of Russia, particularly Siberia. Temperatures arrived at the midpoint of over Russia for January to August were 3.7°C better than expected, 1.5°C over the past record set in 2007. In pieces of northern Siberia, the year to date has been 5°C or more better than expected. The intensity finished in late June, when it came to 38.0°C at Verkhoyansk on the twentieth, temporarily the most noteworthy known temperature anyplace north of the Arctic Circle. Unusual warmth additionally stretched out to different pieces of the great Arctic external Russia, with temperatures arriving at records on 25 July of 21.9°C at Eureka (Canada), and 21.7°C at Svalbard Airport. Broad rapidly spreading fires happened through many pieces of northern Siberia, adding to record high fierce blaze related fossil fuel byproducts during the Arctic summer and the glow added to extraordinarily early ocean ice retreat along the Russian Arctic coast.

Various outstandingly huge fierce blazes, including the biggest flames at any point kept in the territories of California and Colorado, happened in the western United States in pre-fall and harvest time.

Far reaching dry spell conditions through the western portion of the country, especially the inside southwest, added to the flames, as did an extremely powerless summer rainstorm. July to September were the most sultry and driest on record for the southwest. Strange lightning movement in seaside California in mid-August likewise touched off many flames. The most horrendous flames were in California and western Oregon, with more than 8,500 designs obliterated in California and 2,000 in Oregon; 41 passings altogether were ascribed to the flames across numerous states. There were likewise various episodes of outrageous intensity. Passing Valley came to 54.4°C on 16 August, the most elevated known temperature on the planet is essentially the most recent 80 years, while 49.4°C on 6 September at Woodland Hills was a record for more noteworthy Los Angeles.

Significant fierce blazes in eastern Australia which had consumed the later piece of 2019 went on into mid 2020, preceding at last being controlled after weighty downpour toward the beginning of February. Dry spell conditions which had won since mid 2017 facilitated from January onwards, yet there were various episodes of outrageous intensity in mid 2020. Penrith in western Sydney came to 48.9°C on 4 January, the most noteworthy saw in an Australian metropolitan region, while Canberra, which set month to month standards in each of the three mid-year months, arrived at another high of 44.0°C around the same time. Serious smoke contamination additionally impacted many pieces of southeastern Australia in the early piece of 2020. Various stations in New Zealand announced their longest drought on record between late December 2019 and late February 2020.

A significant heatwave impacted the Caribbean district and Mexico in April. Temperatures came to 39.7°C at Veguitas on 12 April, a public record for Cuba, while Havana additionally had its most smoking day with 38.5°C. In eastern Mexico temperatures surpassed 45°C at various areas, coming to as high as 48.8°C, at Gallinas on 12 April, while exceptionally high readings in Central America included 41.2°C at San Agustin Acasaguatlan (Guatemala). Further outrageous intensity in September saw public or regional records set for Dominica, Grenada, and Puerto Rico.

Dry circumstances impacted pieces of north-focal Europe during spring and summer 2020, albeit by and large not similarly as in 2018 or 2019. April was particularly dry, with Romania and Belarus having their driest April on record and Germany and the Czech Republic their second driest, while Geneva, Switzerland had a record 43-day drought from 13 March to 24 April. There was additionally a critical heatwave in western Europe in August. While it was by and large not quite as extreme as those of 2019 (with the exception of locally on the northern shore of France), numerous areas, especially in northern France, arrived at temperatures which positioned second behind the 2019 heatwaves, and De Bilt (Netherlands) had a record eight continuous days over 30°C. Toward the beginning of September the focal point of outrageous intensity moved toward the eastern Mediterranean, with all-time records at areas including Jerusalem (42.7°C) and Eilat (48.9°C) on 4 September, following a late July heatwave in the Middle East wherein Kuwait Airport came to 52.1°C and Baghdad 51.8°C.

It was an exceptionally sweltering summer in pieces of east Asia. Hamamatsu (41.1°C) equaled Japan's public record on 17 August, and Taipei had its most smoking day on record with 39.7°C on 24 July. Hong

Kong had a record run of 13 successive hot evenings, with an everyday least temperature of 28°C or above, from 19 June to 1 July, trailed by 11 continuous sweltering evenings from 5 to 15 July.

Long haul dry spell kept on persevering in pieces of southern Africa, especially the Northern and Eastern Cape Provinces of South Africa, albeit weighty winter downpours saw water stockpiles arrive at full limit in Cape Town, proceeding with the recuperation from outrageous dry season, which topped in 2018.

Precipitation during the 2019–20 summer stormy season in inside southern Africa was locally weighty, yet long haul dry spell endured in certain areas.

6.17.3. Extreme Cold and Snow

North America's most critical blizzard of the 2019–20 winter happened on 17–18 January in Newfoundland. St. John's gotten 75 cm of snow, including a record everyday snowfall, and wind whirlwinds km/h. Later in the year, there were two outrageous early-season cold episodes in pre-winter. In the second seven day stretch of September, boundless snowfalls happened in marsh Colorado, including Denver, where a September record high of 38.3°C had been set just three days sooner, on 5 September.

Afterward, in October, a significant cold flare-up brought outstandingly low temperatures and winter precipitation across a wide locale of the Rocky Mountains and focal states. A harming ice storm in Oklahoma City saw blackouts which went on for a really long time across the greater part the city, while additional north, Potomac, Montana came to −33.9°C on 25 October, the earliest pre-winter date on which temperatures had decreased underneath −30°C at an environment station anyplace in the United States (barring Alaska).

The incredibly wet and warm winter in northern Europe brought about especially low snow cover in many spots–Helsinki encountered a record low number of snow-covered days, breaking the past record by a long shot–yet in far northern Europe, where temperatures were better than expected yet at the same time cool enough for snow, the snowpack was particularly weighty. At Sodankylä (Finland), the snowpack arrived at record profundities in mid-April. A virus May prompted a deferred liquefy with some cover persevering into June, while flooding happened in late May and early June with the spring melt. Downpour on-snow occasions forestalled some reindeer groups from arriving at feed.

It was a virus winter in southern South America. Tierra del Fuego had its most huge virus spell beginning around 1995 in late June and early July, with Rio Grande recording a limit of −8.8°C and least of −16.5°C on 1 July. Paraguay saw record least temperatures for August in various spots. Snow cover in Patagonia was the second-generally broad starting around 2000 and ocean ice framed along parts of the Tierra del Fuego coast. A few stock misfortunes were accounted for. In August, a virus wave in the north of Peru's Amazonia saw temperatures at Caballacocha arrive at 12.8°C, the least temperature recorded there starting around 1975.

Unusual low-rise snowfalls happened in Tasmania toward the beginning of August. On 4 August, snow settled to the ocean level in Launceston, the city's most huge snowfall beginning around 1921. Liawenee, in the focal good countries, came to −14.2°C on 7 August, a Tasmanian record low.

6.17.4. Tropical Storms and an Exceptionally Active North Atlantic Hurricane Season

The quantity of hurricanes around the world was better than expected in 2020, with 96 named typhoons (to 17 November) in the 2020 Northern Hemisphere and 2019–20 Southern Hemisphere seasons.

The North Atlantic area had an exceptionally dynamic season, with 30 typhoons starting around 17 November, over two times the drawn out normal and breaking the record for a full season, set in 2005. Most different bowls had tornado numbers close or somewhat less than ideal.

The Accumulated Cyclone Energy (ACE) record, which incorporates twister power and life span, was well less than ideal in all bowls aside from the North Atlantic and North Indian Oceans, with the Northwest Pacific around half underneath the drawn out normal. While ACE occasional qualities for the North Atlantic were better than expected, they were well shy of occasional records.

The outstandingly dynamic North Atlantic season brought about an enormous number of landfalls. 12 frameworks made landfall in the United States, breaking the past record of nine; five of them in the province of Louisiana. The most extreme effects of the time in the United States came from Laura, which arrived at classification 4 power and made landfall on 27 August close to Lake Charles in western Louisiana, with broad breeze and tempest flood harm. Laura was additionally connected with broad flood harm in Haiti and the Dominican Republic in its creating stage. 77 passings and US$14 billion in financial losses were ascribed to the tempest across the three nations. Later in the season, two significant typhoons made landfall in fast progression in Central America. Storm Eta made landfall as a class 4 framework on the east shoreline of Nicaragua on 3 November, causing serious flooding in the locale as it moved gradually across Nicaragua, Honduras, and Guatemala. Estimated time of arrival moved seaward on 6 November and once again increased to a typhoon, making further landfalls in Cuba and on the Florida Keys. Particle was much more grounded, turning into the season's most memorable classification 5 framework off the Nicaraguan coast on 16 November.

Other harming landfalls in the United States remembered those of Isaias for August in North Carolina and Sally in September in Alabama, while different pieces of the area to encounter critical harm during the season included Bermuda, the Bahamas and the Yucatan Peninsula of Mexico.

Typhoon Amphan, which made landfall on 20 May approach the India-Bangladesh line in the eastern Bay of Bengal, was the costliest hurricane on record for the North Indian Ocean with revealed monetary misfortunes in India of roughly US$14 billion. Huge scope departures of beachfront regions in India and Bangladesh implied that setbacks were far lower than in past similar tornadoes in the area. By the by, 129 lives were lost across the two nations. Amphan arrived at class 5 comparable force while over the Bay of Bengal and, while it debilitated fairly before landfall as a classification 2 tempest, it actually prompted broad breeze and tempest flood harm in the city of Kolkata and encompasses.

Harold, in the Southwest Pacific in April was another class 5 twister. Its most huge effects were in the northern islands of Vanuatu, where it made landfall at close to top power on 6 April.

In excess of 17 000 homes were harmed or destroyed, and around 65% of Vanuatu's populace was affected, with the city of Luganville among the hardest hit. There was likewise huge harm in Fiji as Harold passed toward the south of the primary island of Viti Levu, while harm additionally happened in Tonga and the Solomon Islands. passings were accounted for along the twister's way, the greater part of them when high oceans washed travelers over the edge from a ship in the Solomon Islands.

The most grounded hurricane of the time was Typhoon Goni (Rolly). It crossed the northern Philippines on 1 November with a 10-minute mean breeze speed of 220 km/h (or higher) at its underlying landfall, making it one of the most serious landfalls on record. This was followed the following week by Vamco (Ulysses), which was less serious but caused extreme flooding in Manila and different regions. Something like 25 passings were credited to Goni (Rolly) and 67 to Vamco (Ulysses)51. Two significant hurricanes hit the Korean Peninsula inside a couple of days toward the beginning of September, with Typhoon Mayask (Julian) making landfall close to Busan on 3 September, trailed by Haishen (Kristine) on the seventh. Mayask (Julian) brought 963.5mm of precipitation north of two days to a site on Jeju Island, and wrap blasts on the island up to 177 km/h. Both typhoons prompted huge flooding on the Korean Peninsula and in western Japan, and 41 lives were lost when a boat sank off western Japan during the entry of Maysak. Eight hurricanes and discouragements made landfall in Vietnam in under five weeks in October and November, adding to significant flooding in the district (see Floods area). Other typhoons related with huge death toll through flooding were Cyclone Diane in Madagascar in January, and Tropical Storm Amanda in Guatemala and El Salvador in May.

6.17.5. Extratropical Cyclones and Severe Local Storms

One of the costliest outrageous occasions of it was a derecho (quick line of extreme tempests) which got across the U.S. Midwest on 10 August, stretching out from South Dakota to Ohio. Iowa was the most awful impacted region, with extreme rural misfortunes, including an expected 2 million hectares of corn and soybean crops smoothed. More than 90% of the structures in the city of Cedar Rapids were harmed. Complete misfortunes were assessed at US$7.5 billion. The U.S. cyclone season to date has been close to average, with the two most huge flare-ups influencing the Nashville region on 2–3 March and a wide region stretching out from Texas to Maryland on 12–13 April. 25 lives were lost in the March episode and 35 in the April flare-up, with misfortunes of a few billion bucks across the two events.

Serious tempests impacted a wide area of eastern Australia in the third seven day stretch of January. The main harm was in Canberra, where hail five cm in breadth on 20 January prompted broad vehicle and building harm across a wide wrap of the city. Huge hail harm likewise happened in a few eastern rural areas of Melbourne. Complete misfortunes from the tempests were assessed at US$1.1 billion. Further significant hailstorms impacted Queensland on 31 October with hailstones as extensive as 14 cm in measurement announced in suburbia of Brisbane. Other major serious tempests incorporated a gust line which crossed southern Brazil on 30 June, with cyclones, hail, and wind blasts surpassing 100 km/h, with 12 passings revealed, and a hailstorm in Tripoli (Libya) on 27 October with hailstones as extensive as 20 cm.

Various serious extratropical tornadoes impacted northern, focal, and western Europe in February, adding to an exceptionally wet month with flooding in some areas. The main tempest of this period was Ciara (otherwise called Sabine and Elsa), which passed toward the northwest of Europe on 10–11 February. Critical breeze harm reached out from Ireland to Sweden, Poland, and Italy. In the pre-winter, Storm Alex toward the beginning of October carried outrageous breezes to western France with blasts as high as 186 km/h, while weighty downpour reached out across a wide region. 3 October was the wettest region found the middle value of day on record for the United Kingdom with a public normal of 31.7 mm, while outrageous precipitation happened close to the Mediterranean coast on the two sides of the France-Italy line, with 24-hour sums surpassing 600 mm in Italy and 500 mm in France. The outrageous precipitation reached out to southern Switzerland with 421 mm in 24 hours at Càmedo. Huge blaze flooding happened in each of the three nations.

A Mediterranean twister cross breed framework ('Medicane'), showing a few attributes of a typhoon and others of a mid-scope storm, created in mid-September, making landfall in southwestern Greece on the eighteenth with wind blasts surpassing 110 km/h. Prior, it carried weighty precipitation and glimmer flooding to waterfront Libya, with an everyday precipitation complete of 150.5 mm at Sabratha on 15 September.

6.18. SIDEBAR: CLIMATE INDICATORS AND THE SUSTAINABLE DEVELOPMENT GOALS

The 2030 Agenda for Sustainable Development gives a common diagram to harmony and flourishing for individuals and the planet, presently, and into what's in store. At its heart are a bunch of reasonable advancement objectives (SDGs), which perceive that finishing neediness and different hardships should remain closely connected with methodologies that further develop wellbeing and instruction, diminish disparity, and spike monetary development, while handling environmental change and saving seas and woodlands. The accomplishment of a large number of these objectives is seriously jeopardized by environmental change, nonetheless. For instance, increasing temperatures are prompting the deficiency of species and biological systems, which can diminish agrarian and fishing yields—adding to food weakness and influencing vocations (SDGs 1, 2, 14, and 15). Outrageous climate and environment occasions can cause wellbeing risks, harm foundation and lead to water shortage (SDGs 1, 3, 9, 11 and

6). These dangers, along with others, are between related with struggle and dependability (SDG 16). It is basic to feature that these dangers won't affect all populaces or areas similarly and can support or demolish existing imbalances (SDG 10).

The figure underneath exhibits how rising barometrical CO_2 fixations lead to flowing impacts by means of six of the other key environment pointers. Past presenting dangers to accomplishing practical development, some of these cycles likewise can possibly deliver further ozone-depleting substances into the climate in a criticism circle that can sustain warming. For instance, increasing temperatures can defrost permafrost, delivering more carbon into the climate.

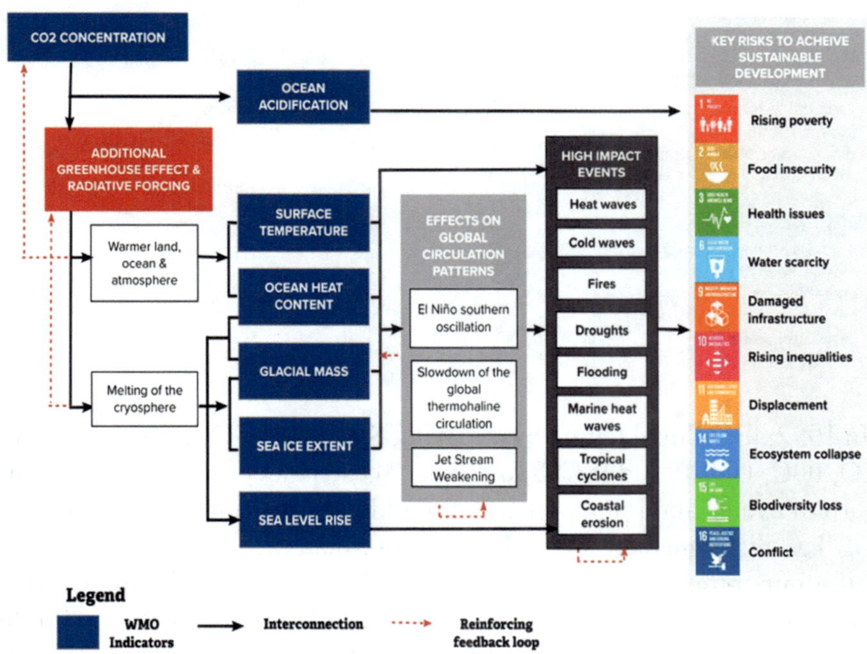

Figure 6 14: Selected climate change-related risks to the SDGs (ICRC, 2020; IPCC, 2014; Trewin, et al; WMO Statement on the State of the Global Climate in 2019., 2020; United Nations, 2015).

6.19. SIDEBAR: GLOBAL CLIMATE OBSERVING SYSTEM

Environment checking is performed by an arrangement of frameworks covering air, seas, hydrology, the cryosphere and biosphere. Every one of these areas is checked in various ways by a scope of associations. Cutting across this multitude of regions, satellite perceptions are a significant supporter of worldwide environment perceptions.

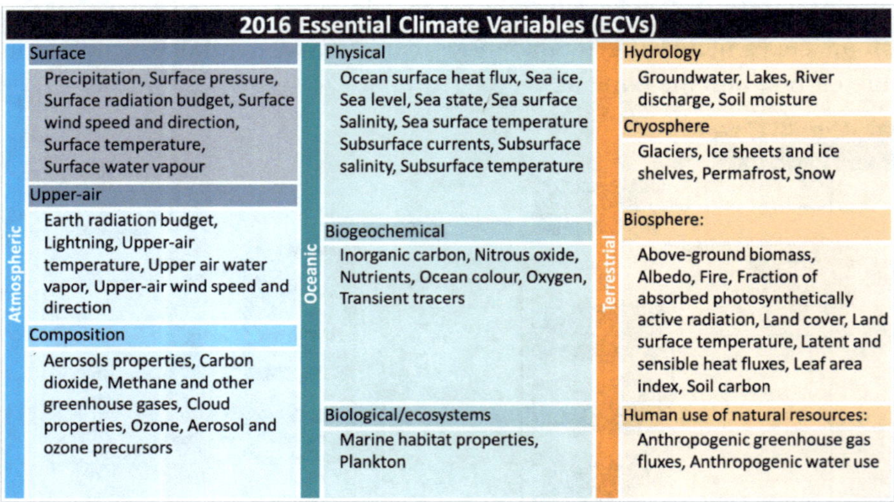

In 1992, the Global Climate Observing System (GCOS) was laid out by WMO, IOC, UNEP, and ISC55 to arrange these perceptions and work with the turn of events and improvement of worldwide environment perceptions. GCOS has distinguished a bunch of factors that, when noticed, ought to give the important data to figure out, model, and foresee the direction of the environment as well as plan relief and variation. The situation with the perceptions of these factors, called Essential Climate Variables (ECV), are checked by GCOS, which distributes ordinary status reports. GCOS likewise distinguishes what is expected to work on the framework in Implementation reports.

GCOS manages three boards for the air, sea, and earthly frameworks. These boards are together supported with the World Climate Research Program (WCRP), and the Global Ocean Observing System (GOOS), for the seas.

While the GCOS organizations, GSN, and GUAN, give a standard organization estimating meteorological factors, National Meteorological and Hydrological Monitoring Services (NMHS), composed by WMO, give a more extensive and far and wide organization of estimations. Additionally, WMO's Global Atmospheric Watch (GAW) facilitates climatic arrangement estimations, guaranteeing dependable and precise information, from estimations made by WMO Members, research organizations as well as offices and other contributing organizations.

Sea perceptions of sea physical science, geobiochemistry science and biological systems are composed through GOOS. The Ocean Observations Physics and Climate Panel (OOPC) together supported by GCOS, GOOS, and WCRP, distinguishes maritime ECV and integrates them into sea network plan. The GOOS Observations Coordination Group (OCG) screens the exhibition of these networks and produces a yearly Ocean Observing System Report Card.

In the earthbound area there is a more extensive gathering of noticing networks. Hydrological perceptions are for the most part worked by NMHS and facilitated through WMO. There are various Global Terrestrial Networks (GTN) that report to GCOS. These incorporate the GTN-Hydrology (GTN-H) that itself is a cooperation of different organizations like those covering lakes, waterways, groundwater, soil moisture, and water use. The GTN-Permafrost (GTN-P) GTN-Glaciers (GTN-G) with the World Glacier Monitoring Service (WGMS) are progressively helping out the WMO Global Cryosphere Watch (GCW). GOFC-GOLD covers over the ground biomass, woodlands, land use, and fierce blazes and predominantly gives direction and alignments to the for the most part satellite-based perceptions that are expected to cover biological systems across the globe.

Notwithstanding this there are various locally based networks that are critical supporters of the organizations: some are broadly based while others cover more extensive districts.

The CEOS/CGMS Joint Working Group on Climate (WGClimate) bases the improvement of satellite perceptions for environment on the ECV necessities laid out by GCOS. They have delivered an ECV Inventory that incorporates records for 766 environment information records for 33 ECV covering 72 separate ECV items with more arranged. Satellite perceptions enjoy a few benefits, they have a close worldwide inclusion, yet optical perceptions can be interfered with by mists. Utilized with ground-based perceptions, either as correlative informational indexes, or for approval and adjustment they structure a significant piece of the worldwide noticing framework.

6.20. RISKS AND IMPACTS

The risk of environment related influences relies upon complex associations between environment related dangers and the weakness, openness, and versatile limit of human and regular frameworks. At current degrees of worldwide ozone-depleting substance discharges, the world stays on track to surpass the concurred temperature limits of either 1.5°C or 2°C above pre-modern levels, which would expand the risk of unavoidable impacts of environmental change past what is as of now seen.

Environment related occasions as of now present dangers to society through influences on wellbeing, food, and water security as well as human security, livelihoods, economies, framework, and biodiversity. Environmental change likewise has suggestions for biological system administrations. It can influence examples of normal asset use, as well as the conveyance of assets across areas and inside nations.

Environmental change and individual occasions have an enormous natural impact also. Negative natural impacts remember influences for the land like dry seasons, rapidly spreading fires in woodland and peatland regions, land debasement, sand, and residue tempests, and desertification. Air contamination is connected to the utilization of non-renewable energy sources. In freshwater frameworks, floods, and water pressure, and in marine frameworks an assortment of effects including ocean level ascent, sea fermentation, diminished degrees of sea oxygen, mangrove rot and coral dying. A significant number of these effects are connected to biodiversity misfortune. For instance, other than the weighty human effects of the fierce blazes in Australia in 2019 and mid 2020, there were probable serious misfortune or uprooting of a large number of animals during the disaster.

6.20.1. Human Mobility and Displacement

Environment and climate occasions differently affect populace developments and on the weakness of individuals moving consistently. In accordance with recently detailed designs, a considerable lot of the biggest scale removals were in crowded Asian nations. Besides, the intricacy of human-portability related chance and effects will in general be most noteworthy–and least comprehended because of significant impediments in the accessible information–comparable to sluggish beginning occasions and in delicate and struggle impacted settings. Information in this space stays pertinent for worldwide and public strategy improvement, tasks, and exploration in the areas of human versatility and environment activity. In 2020, the COVID19 pandemic has added a further aspect to human portability concerns featuring the requirement for a coordinated way to deal with understanding and tending to environment chance and effects on weak populaces.

6.20.1.1. Weather and Climate-Related Hazards and Disasters Driving New Displacement

Throughout the last 10 years (2010–2019), climate related occasions have set off an expected 23.1 million relocations of individuals on normal each year. By far, most of these developments happen inside public boundaries, however cross-line developments may likewise happen. A few 9.8 million removals, generally due to hydro-meteorological dangers and debacles, were recorded during the principal half of 2020, primarily moved in South and South-east Asia and the Horn of Africa. Occasions in the final part of the year, including removals connected to flooding across the Sahel area, the dynamic Atlantic storm season and hurricane influences in Southeast Asia, are supposed to bring the complete for the year near the normal for the ten years.

In May, Cyclone Amphan hit the Sundarbans locale among India and Bangladesh. 2.4 million individuals were uprooted in India, for the most part in West Bengal and Odisha, and 2.5 million in Bangladesh. While many returned generally soon a short time later, harm to more than 2.8 million homes probably brought about vagrancy and delayed dislodging for some thousands. Many uprooted individuals didn't approach clearing focuses and were constrained to take cover in tents or in the outside on embankments.

Storm downpours prompted huge flooding, dam breakdowns, avalanches, and the relocation of millions of individuals in Bangladesh, China, India, Japan, Pakistan, Nepal, South Korea, Turkey, and Vietnam. The floods and avalanches in China were especially extreme. 29 000 homes were obliterated and over 2.2. million individuals cleared by mid-July 2020.

In July, flooding, and avalanches impacted a few locale across Nepal. IOM evaluations in August showed the presence of a sum of 5 467 people from 1 066 families in 29 dislodging locales across the country. By September, just 12 locales were as yet dynamic, facilitating around 2 000 people in three areas, as individuals got back with the finish of the rainstorm season.

In the Philippines, Super Typhoon Goni (privately known as Rolly) uprooted 400,000 when it made landfall across the Philippines on 1 November. After fourteen days, Typhoon Vamco set off a further 320 000 displacements. In October 2020, Cyclone Molave, the fourth tempest of the month to hit Vietnam, set off the departure of a few 1.3 million individuals and left countless individuals in a basic circumstance as harvests were obliterated.

6.20.1.2. Patterns of Protracted, Prolonged, and Repeated Displacement Linked with Hydro- Meteorological Hazards

2020 likewise saw probably the biggest fierce blazes on record, which prompted critical uprooting. The 2020 fire season in the USA brought about the annihilation of something like 2 000 private structures in California and dislodged around 100 000 individuals. In Oregon, around 500 000 individuals were set under clearing notice and more than 40 000 must be emptied from their homes. In Australia, around 65 000 new relocations were recorded as a result of the flames that impacted the country between July 2019 and February 2020. Fires additionally obliterated 3 100 homes, possibly prompting longer-term relocation of around 8 100 individuals.

6.20.1.3. Compounded Risk and Vulnerability

Exiles, inside dislodged individuals and transients are frequently among those generally powerless against environment and climate related dangers. Numerous weak individuals moving, no matter what their explanations behind moving, wind up getting comfortable high risk regions, where they are presented to environment and weather conditions dangers at a scope of scales. Weather conditions perils and human versatility may likewise converge with social and political pressures and struggle in complex settings and, in that capacity, require the coordinated thought of multi-danger calamity risk decrease measures, including early admonition frameworks and readiness, and longer-term reasonable improvement concerns, for example, land use and metropolitan preparation.

Explicit states of chance from hydro-meteorological disasters are especially irrefutable in the Rohingya evacuee locales in Cox's Bazar, Bangladesh. Over the course of the year, an aggregate of 162 275 have been impacted, with many requiring explicit assistance. Without readiness estimates attempted in the camp regions, including units for reinforcing covers, working of holding structures on slopes, and further developing seepage, streets, and scaffolds, these effects would have been more regrettable.

In East Africa, floods, and the most exceedingly terrible desert beetle pervasion in 25 years, brought about huge scope relocation. In Somalia, floods were related with the removal of more than 1,000,000 individuals in 2020, for the most part inside the country, while dry season related influences prompted a further 80,000 displacements. Around a fourth of the populace impacted by floods this year were individuals previously living in packed and shaky camps for inside uprooted people81, to which many recently flood-dislodged individuals likewise moved. Ranchers whose harvests were crushed by desert grasshoppers were likewise compelled to move looking for endurance assistance.

In Ethiopia, north of 300 000 removals were welcomed on by floods and avalanches during the principal half of the year, 66% of which were in the Somali region. IOM evaluations among June and July 2020 have shown the presence of somewhere around 104 696 individuals uprooted by floods and of 351 062 by dry seasons in the country's South-East–separately 6% and 19% of the nearby dislodged populace.

Sudan experienced streak floods and waterways, including the Nile, flooded in August. Nearly 125 000 outcasts and inside uprooted people were among those affected.

Among August and September 2020, outrageous precipitation set off floods that struck a few nations in the Sahel. IOM appraisals assessed an aggregate of 1.25 million individuals uprooted in Burkina Faso, Mali, and Niger–in settings previously portrayed by struggle and food uncertainty, as well as constrained interior and cross-line populace developments. The further corruption of day to day environments and livelihoods coming about because of the floods expanded the risk of dislodging inside the area, toward the south to waterfront nations and toward the north to North Africa and Europe. Floods in August 2020 in N'Djamena, Chad, brought about 31,853 people escaping their homes and tracking down cover somewhere else in the city, with little alleviation help being given. Floods additionally impacted the Lac area, dislodging north of 4,000 individuals, and influencing a lot more who had recently been uprooted by struggle and violence. In Yemen, where over 80% of the complete populace is needing compassionate help, serious blaze floods uprooted numerous who were at that point interior dislodged due to the contention, residing in states of wretched destitution, in unsafe asylums and without admittance to get, elective spots in which to shield. Nearly 300 000 individuals lost their homes, harvests, domesticated animals and possessions because of downpours and floods. The abrupt breaking of the

Al-Roone dam impacted a large number of individuals in IDP destinations in Al-Tahseen, Souq al-Lill and elsewhere. In Syria from early October, out of control fires in Lattakia, Tartous, and Homs impacted up to 40 000 individuals through harm to lodging and resources and significant loss of livelihoods. About 25 000 removals were accounted for, incorporating from regions with as of late brought individual development plans (IDPs) back. The strength of impacted networks has been dissolved more than 10 years of contention, as well as by the COVID-19 pandemic, monetary emergency and the first monstrous take-up of uprooted individuals.

6.20.1.4. COVID-19 Impacts

Versatility limitations and financial slumps because of COVID-19 have eased back the conveyance of philanthropic help to weak individuals progressing as well as endeavors to help recuperation for impacted people, including sturdy answers for those uprooted. The weaknesses of uprooted populaces, who frequently live in thickly populated settlements, were intensified further. Typhoon Harold, probably the most grounded storm at any point kept in the South Pacific, which struck Fiji, the Solomon Islands, Tonga, and Vanuatu set off an expected 99 500 removals. In view of COVID-19 lockdowns and measured isolations, reaction, and recuperation tasks were hampered prompting delays in giving hardware and assistance. Moreover, in the Philippines, the pandemic confounded departure and reaction endeavors in front of Tropical Cyclone Vongfong (Ambo) in mid-May. More than 180 000 individuals were prudently evacuated, through activities were hampered by the requirement for social separating measures as inhabitants couldn't be shipped en masse and clearing focuses must be utilized at half capacity. The tempest likewise harmed Bicol Region's just COVID-19 testing facility. In northern Central America, a few 5.3 million individuals needed philanthropic help including 560 000 inside uprooted individuals before the beginning of the pandemic. Reactions to tempests and typhoons that over and again impacted the region all through the Atlantic storm season have consequently occurred in a setting of complex, interlinked weakness, connected to pre-COVID-19 and COVID-19-related difficulties. Making the circumstances that empower individuals' choices to securely remain, to use the maximum capacity advantages of noble relocation, and to turn away, limit, and address removal and related assurance needs is a critical worry for both environment and human portability arrangements and participation–particularly as nations are confronted with choices to grow versatility choices that have been compelled all through the COVID-19 emergency.

6.20.2. Climate-Related Hazards and Compounded Crisis Impacts on Food Security

Following quite a while of decline, the new expansion in food frailty starting around 2014 is driven by struggle, monetary lull as well as environment inconstancy and outrageous climate occasions. Almost 690 million individuals or 9% of the total populace were undernourished and around 750 million–or almost 10% – were presented to extreme degrees of food weakness in 201997. The quantity of individuals ordered under emergency, crisis, and starvation conditions (Integrated Food Security Phase Classification 3 or above) had expanded to very nearly 135 million individuals across 55 countries. In 2020, 12 out of the 20 most weak nations to environmental change–and with the least versatile limit–were in conflict. The world has likewise seen 33.4 million new inner relocations across 145 nations and regions in 2019, set off by complex financial weaknesses and, frequently simultaneously happening, natural corruption and environment related disasters. On top of these intensified drivers of worldwide food instability, the COVID-19 pandemic has added new difficulties for food unreliable individuals and food esteem chains all over the planet.

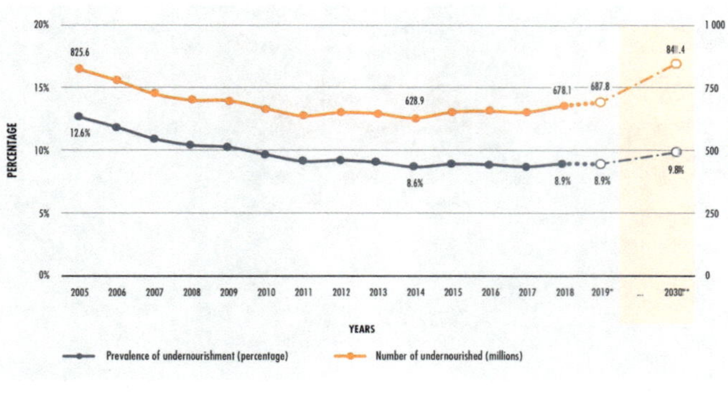

Figure 6.15: The number of undernourished people in the world continued to increase in 2019. Projected values in the figure are illustrated by dotted lines and empty circles. The shaded area represents projections for the longer period from 2019 to the 2030 target year. The entire series was revised to reflect new information made available since the publication of the last edition of the report; it replaces all series published previously.

Source: FAO.

In 2020, more than 50 million individuals have been doubly hit: by environment related debacles (floods, dry spells, and tempests) and the COVID-19 pandemic. Nations in Central America are experiencing the triple-effect of tropical storms Eta and Iota, COVID-19 and previous philanthropic emergencies. The Government of Honduras assessed that 2.8 million individuals were impacted by typhoon Eta, and 53 000 hectares of cropland were washed away, principally rice, beans, and sugarcane. Among the Pacific Islands, hurricane Harold has impacted north of 17 500 hectares of cropland in Vanuatu, causing significant harm similarly as the developing season began. Weighty downpours in the Arabian Peninsula and East Africa brought about the biggest desert beetle flare-up in 25 years across the Horn of Africa. In Ethiopia alone, 200 000 hectares of cropland were harmed and more than 356 000 tons of grains were lost, leaving very nearly 1,000,000 individuals food insecure. The significant stretch of dry weather conditions previously saw in South America in 2020 is probably going to go on as La Niña frequently carries sub optimal precipitation to certain areas. Accordingly, corn creation in Argentina is conjecture to drop by 2 million tons (5% decrease) in 2020–2021.

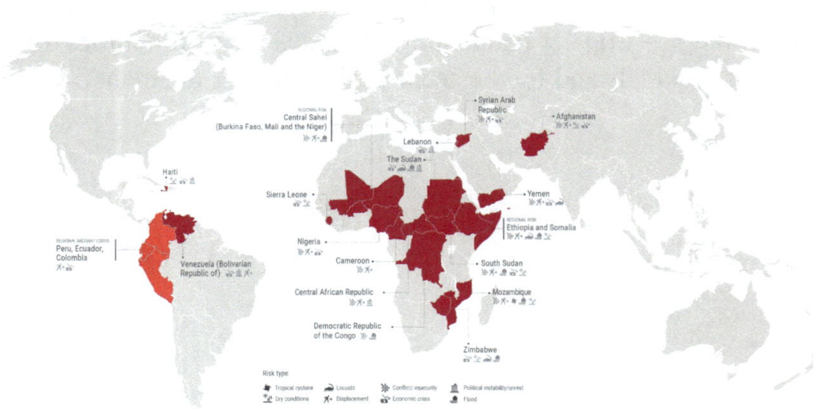

Figure 6.16: Displays FAO-WFP early warning analysis of acute food insecurity hotspots in July 2020 [107]. The hotspot countries were selected through a consensus-based analytical process by FAO and WFP, focusing primarily on: levels of food insecurity and malnutrition prior to the COVID-19 pandemic, projections of acute food insecurity, presence of existing compounding shocks and stressors, including climatic conditions, plant pests and animal diseases, economic crisis or political instability, conflict, and insecurity.

The limitations to forestall the spread of COVID-19, especially measures to restrict the development of individuals inside and among nations and the deal and acquisition of labor and products, have presented critical difficulties for dealing with the effects of environment related calamities on food frailty.

The attainability of eye to eye administrations to furnish ranchers with data from occasional weather conditions estimates as direction on early activities has been diminished in numerous countries.

Interruptions of the farming area by COVID-19 have exacerbated weather conditions influences along the whole food store network, lifting levels of food instability, hunger, and undernourishment. Specifically, the pandemic limitations on development of workforce and interruption of the inventory network of hardware has made new difficulties to battle desert beetle flare-ups brought about by weighty downpour in the Greater Horn of Africa and Yemen, where many undernourished individuals are right now found. Another model in Somalia. By September 2020, information recommended that the quantity of Somalis confronting intense unhealthiness (IPC Phase 3 or above) and imperceptibly ready to meet least food needs significantly increased to 3.5 million contrasted with mid 2020109. Desert insects and stream floods were liable for crop misfortunes inside the country with an expected decrease in oat creation of 15–25% during the April-June developing season.

FAO's gauges and following of beetle swarms in the Sindh area in Pakistan, joined with nearby circumstance examination, has assisted the Government of Pakistan with setting up successful reconnaissance and control activities. Therefore, Pakistan had the option to review 61.4 million hectares and treat 1.13 million hectares of cropland now pronounced as beetle free zones.

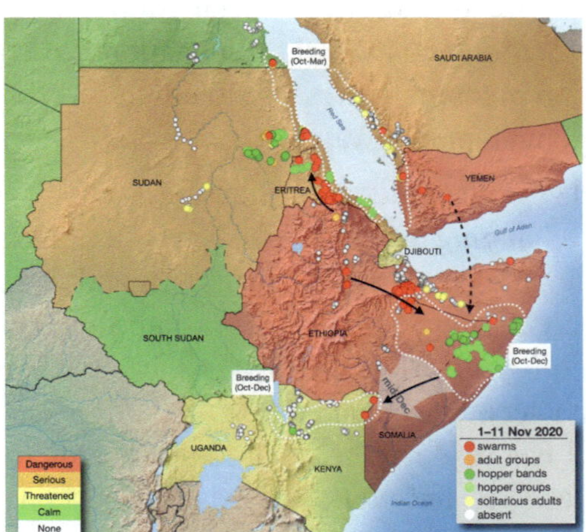

Figure 6.17: Desert locust movement prediction using FAO´s Desert Locust Information System (DLIS). The unusual weather conditions in fall 2019 along

the Arabian Peninsula generated strong cyclones and heavy rains resulting in a higher than normal vegetation growth providing ideal conditions for desert locusts. During the 2020 compounded crisis, early warning and early action response played a key role in saving up to 1.52 million tons of cereals (enough to feed 10 million people in one year), worth around US$456 million, across nine eastern African countries and Yemen. Other anticipatory actions promoted by the FAO include: i) providing key inputs to the most food-insecure farmers, including seeds and agricultural tools to allow timely planting; ii) implementing cash-based programs; iii) distributing animal feed and mineral supplements to vulnerable livestock raisers; iv) monitoring and controlling all infested/breeding areas, enhancing surveillance and conducting control operations to prevent swarms from forming and spreading..

6.21. SIDEBAR: CLEAN ENERGY, DISASTER-PROOF INFRASTRUCTURE AND EARLY WARNING SYSTEMS

Meeting the point of the 2015 Paris Agreement "to fortify the worldwide reaction to the danger of environmental change by keeping a worldwide temperature climb this century well under 2°C above pre-modern levels and to seek after endeavors to restrict the temperature increment considerably further to 1.5°C," requires sizeable decreases in fossil fuel byproducts. The International Monetary Fund in the October 2020 World Economic Outlook shows that a push in green framework speculation joined with consistently rising carbon costs could convey the emanation decreases expected to restrict the expansion in temperatures and to put the worldwide economy on a more reasonable development way.

Environmental change moderation strategies—, for example, emanation limits, discharge exchanging programs, feed-in levies that give a reliable cost to makers of sustainable power, carbon charges, and R&D sponsorships—have bit by bit become more famous after some time (Figure 6.18a). The presentation of ecological arrangements has been joined by expansions in worldwide advancement and interest in clean energy innovations (Figure 6.18b) and has made significant commitments to redistribution of development, power age, and work towards low-carbon exercises (Figure 6.18c-e).

A further green foundation speculation push and appropriations for environmentally friendly power age, joined with pre-reported bit by bit expanding carbon costs, would prompt outflows decreases that are expected to keep temperature increments to more secure levels (Figure 6.18). Thusly, designated help for the more unfortunate families and approaches that work with work redistribution from high-carbon to low-carbon areas would guarantee a fair progress. The expenses of the progress to a low-carbon worldwide economy are sensible, with green venture helping monetary action and work and subsequently making up for a portion of the expenses related with carbon charges.

While the ongoing worldwide downturn brought about by the COVID-19 pandemic can make it trying to establish the arrangements required for moderation, it likewise presents valuable chances to put the economy on a greener way by giving right cost signals and other monetary motivating forces to help interest into green and strong public framework, hence supporting GDP and work during the recuperation stage.

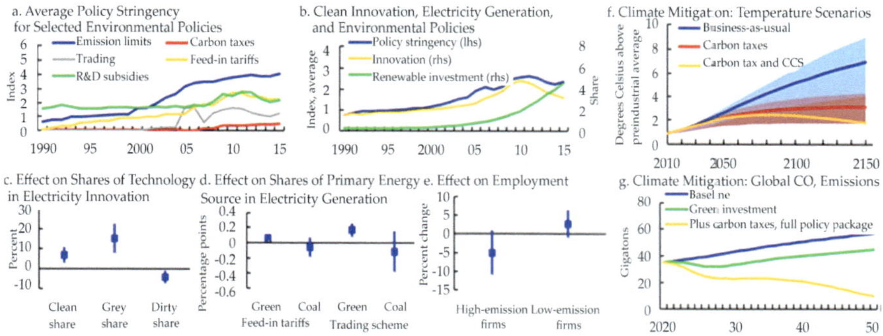

Figure 6.18: Impact of climate change mitigation policies. Sources: IMF (2020b); International Energy Agency; OECD Environmental Policy Stringency Index; and Worldwide Patent Statistical Database. Note: In panels c, d, e whiskers represent 90 percent confidence intervals. In panel f shaded areas represent high and low climate sensitivity. CCS = carbon capture and storage. See IMF (2020b) for full details.

Transformation arrangements pointed toward upgrading flexibility to a changing environment—approaches, for example, interest in resistant to calamity foundation and early admonition frameworks, risk-sharing through monetary business sectors, and advancement of social security nets—can restrict the effect of climate-related shocks and assist the economy with recuperating quicker (IMF 2017). Variation techniques assume a key part in nations that are especially powerless against environmental change, for example, low-pay nations situated in hot districts and regions presented to more successive or extreme catastrophic event risks.

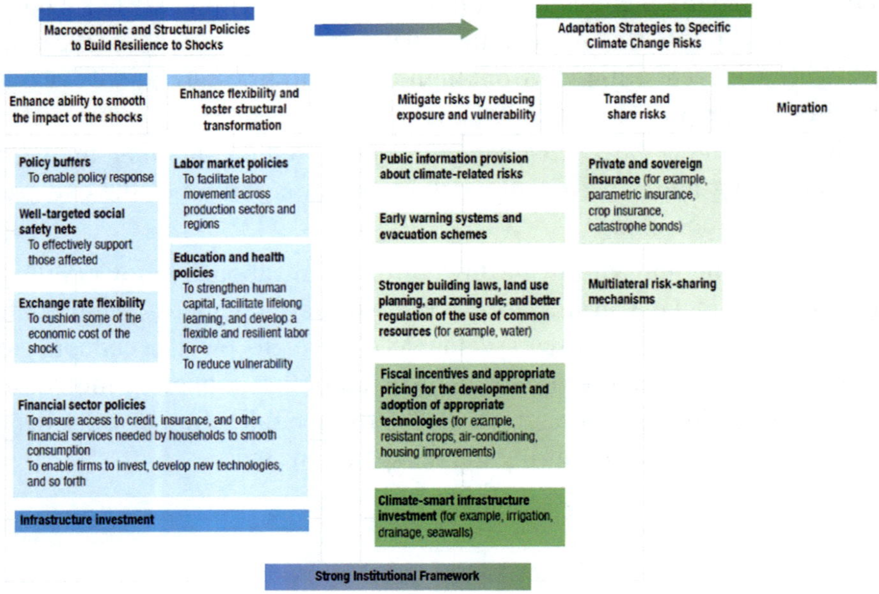

Figure 6.19: Climate change adaptation policies toolkit.

Source IMF (2017).

CHAPTER 7

THE STAKEHOLDER PERSPECTIVE

CONTENTS

7.1. GOVERNMENT

In creating and developed nations, governments are updating strategies in light of key sustainability issues, including ozone-depleting substance discharges, dangerous chemicals, water utilization, work, and human rights.

In the United States, there is restored enthusiasm for growing more compelling oversight of and responsibility for corporate exercises that affect society and the earth. Specifically, the eagerly awaited control of carbon outflows will give a more level playing field for bringing down producers. However, these measures are very much dependent on the elections of 2017.

7.2. INVESTORS

More investors are currently requesting that organizations detail and measure sustainability dangers and opportunities in their monetary divulgences. As organization proprietors, they look to sustainability execution as an indicator of solid administration and long term considering future development potential. Long-term speculators compensate organizations that coordinate sustainability into key objectives. Sustainability-based stock files benchmark corporate execution against peer organizations and place weight on organizations to meet expanding desires.

7.3. WORKER'S UNIONS

Unions anticipate that organizations will address issues that influence employees, regardless of whether they are financial issues, for example, wages, and social insurance, or natural issues, for example, wellbeing, and environmental change. They effectively activate and speak to their individuals through formal arrangements, engagement on open approach, and coordinated effort with different businesses that help basic positions.

7.4. COMMON SOCIETY

NGOs and group bunches anticipate that organizations will address their ecological and social effects. Offering voice to planetary concerns and who and what is to come, and additionally those affected by corporate activity today, these gatherings weight organizations through legitimate activity, open crusades and cooperation to address a full scope of sustainability issues.

7.5. BUSINESS PARTNERS AND SUPPLIERS

with will take after similar principles that they improve the situation incorporating Organizations expect that those they work sustainability into their business. Business-to-business connections in this manner progressively consolidate sustainability models and criteria reflected through changes in Request for Proposals (RFP), and acquisition rules. As clients, organizations are themselves pushing sustainability crosswise over areas and the esteem chain. Gatherings of organizations in the car, clothing, electric utility, hardware, and pharmaceutical areas are among those working together to raise sustainability guidelines over their whole industry supply chains.

7.6. CONSUMERS

A developing section of individual customers are putting their cash where their qualities are, requesting a comprehension of the sustainability effects of the items and administrations that they purchase. Purchaser concerns incorporate the conditions under which items are made, the materials utilized, and post-utilize recyclability. An expanding enthusiasm for dependable sustainability data goes with this pattern, including at the purpose of procurement. The monetary emergency has not dulled buyer sense of duty regarding nature, with 78% of U.S. buyers in a 2009 Cone review showing that they are as likely or more inclined to purchase ecologically dependable items as they were before the emergency.

One of the most grounded powers requesting change originates from inside. Current employees and skilled employment competitors look for work that is important and of verifiable incentive to society, and they are set up to get a lower pay in return for work at a socially capable organization. They search out bosses that have an unmistakable vision for their commitment to an economical worldwide economy, and once inside, hope to impact the heading of corporate sustainability and drive upgrades through their particular obligations.

Similarly, as with the current crumple of the monetary markets–originating from an absence of controls, disappointment of responsibility, and deficient exposure of dangers that speculators were expecting–sustainability and environmental change additionally exhibit broad concealed dangers to which the money-related industry needs to give careful consideration.

Progressively, speculators are perceiving and making a move on administration issues, including natural and social issues that posture material business dangers and that can possibly specifically influence long-term monetary execution. On the other hand, they see significant advantages to be picked up by financing organizations that are proactive on these issues.

The Investor Network on Climate Risk (INCR) is centered around the monetary effects of environmental change. The gathering was built up in 2003 with an enrollment of twelve financial specialists speaking to $600 billion in aggregate resources, and now incorporates more than 80 speculators with $8 trillion in aggregate resources. INCR has been exceptionally dynamic, calling for solid national and worldwide atmosphere strategies, asking the U.S. Securities and Exchange Commission for institutionalized revelation of sustainability dangers, specifically captivating organizations, recording investor resolutions, and making critical interests in clean innovation.

We have seen a blast of sustainability and particularly atmosphere-centered research and speculation items in the previous couple of years, quite a bit of it occurring amidst one of our most noteworthy financial moderate downs. We have additionally observed the number of money-related records that address sustainability widen and extend.

A considerable lot of these records are being overseen by "standard" money-related houses, including the S&P/IFCI Carbon Efficient Index, HSBC Climate Index, Prudential Green Commodities Index, FTSE4Good, and NASDAQ's Global Sustainability 50 Index. Also, speculators are seeing the outcomes. For instance, KLD Research and Analytics (now part of Risk Metrics) propelled a Global Climate 100 Index and posted a 57% return (17% annualized) since its dispatch in 2006.

Notwithstanding remain solitary sustainability lists, money-related businesses are incorporating sustainability into center basic leadership. Various banks, including Citi, Morgan Stanley, and Credit Suisse, have set up a carbon-centered due ingenuity process for any future loaning for coal-let go control and other carbon concentrated undertakings. Bank of America has likewise indicated initiative by setting a particular focus to diminish the rate of ozone-depleting substance outflows in its loaning to the utility business, and by transparency openly that it is utilizing a $20 to $40 per ton cost of carbon in assessing credits.

Socially dependable contributing has advanced from being a "boutique" issue to developing at a quicker pace than the universe of all venture resources under expert administration–growing 18% in the vicinity of 2005 and 2007–while professionally oversaw resources became under 3%. A report by Robeco Investment Management and Booz and Company anticipates that dependable contributing will develop to as much as 20% of aggregate resources under administration by 2015.

The budgetary emergency has additionally brought about an expanded concentrate on the hazard administration forms at money-related organizations, with a push for more noteworthy straightforwardness, all-encompassing danger distinguishing proof and a more drawn out term center. The 658 signatories to the Principles for Responsible Investment, which all in all speak to over $18 trillion of advantages, trust it is in the long-term enthusiasm of their recipients to figure sustainability speculation basic leadership.

A 2009 Report by the Asset Management working gathering of the UNEP Financial Initiative declares that speculators have a guardian obligation to think about sustainability segments and those that don't may confront genuine hazard by opening themselves up to lawful liabilities.

The primary concern for business is this: financial specialists are compensating organizations that see how sustainability issues affect their business and that are executing procedures and activities that will empower them to flourish in a sustainable worldwide economy.

CHAPTER 8

THE EIGHT KEY EXPECTATIONS

CONTENTS

8.1. GOVERNANCE FOR SUSTAINABILITY

1. **Board Oversight:** The Board of Directors will provide oversight and accountability for corporate sustainability strategy and performance. A committee of the board will assume specific responsibility for sustainability oversight within its charter.
2. **Management Accountability:** The CEO and company management–from C-Suite executives to business unit and functional heads–will be responsible for achieving sustainability goals.
3. **Executive Compensation:** Sustainability performance results are a core component of compensation packages and incentive plans for all executives.
4. **Corporate Policies and System:** Companies will embed sustainability considerations into corporate policies and risk management systems to guide day-to-day decision-making.
5. **Public Policy:** Companies will clearly state their position on relevant sustainability public policy issues. Any lobbying will be done transparently and in a manner consistent with sustainability commitments and strategies.

8.2. STAKEHOLDER ENGAGEMENT

1. **Focus Engagement Activities:** Companies will systematically identify a diverse group of stakeholders and regularly engage with them on sustainability risks and opportunities, including materiality analysis.
2. **Stakeholder Dialog:** Companies will engage stakeholders in a manner that is ongoing, in-depth, timely, and involves all appropriate parts of the business. Companies will disclose how they are incorporating stakeholder input into corporate strategy and business decision-making.
3. **Investor Engagement:** Companies will address specific sustainability risks and opportunities during annual meetings, analyst calls, and other investor communications.
4. **C Level Engagement:** Senior executives will participate in stakeholder engagement processes to inform strategy, risk management and enterprise-wide decision-making.

8.3. DISCLOSURE

1. **Standards for Disclosure:** Companies will disclose all relevant sustainability information using the Global Reporting Initiative (GRI) Guidelines as well as additional sector-relevant indicators.
2. **Disclose in Financial Findings:** Companies will disclose material sustainability issues in financial filings.
3. **Scope and Content:** Companies will regularly disclose significant performance data and targets relating to their global direct operations, subsidiaries, joint ventures, products, and supply chain. Disclosure will be balanced, covering challenges as well as positive impacts.
4. **Vehicles for Disclosure:** Companies will release sustainability information through a range of disclosure vehicles, including stand-alone reports, annual reports, financial filings, websites, and social media.
5. **Product Transparency:** Companies will provide verified and standardized sustainability performance information about their products at the point of sale and through other publicly available channels.
6. **Verification and Assurance:** Companies will verify key sustainability performance data to ensure valid results and will have their disclosures reviewed by an independent, credible third party.

8.4. PERFORMANCE

1. **Operations:** Companies will invest the necessary resources to achieve environmental neutrality and to demonstrate respect for human rights in their operations. Companies will measure and improve performance related to GHG emissions, energy efficiency, facilities, and buildings, water, waste, and human rights.
   a. **Greenhouse Gas Emissions and Energy Efficiency:** Companies will reduce greenhouse gas emissions by 25% from their 2005 baseline by 2020: Improving energy efficiency of operations by at least 50%. Reducing electricity demand by at least 15%. Obtaining at least 30% of energy from renewable sources.

b. **Facilities and Buildings:** Companies will ensure that at least 50% of their owned or leased facilities, and all new construction, will meet rigorous green buildings standards. When setting up facilities, companies will follow best practices that incorporate sustainable land-use and smart growth considerations.

c. **Water Management:** Companies will assess water-related impacts and risks and will set targets to improve water use and wastewater discharge, with priority given to operations in water-stressed regions.

d. **Eliminate Waste:** Companies will design (or redesign, as appropriate) manufacturing and business processes as closed-loop systems, reducing toxic air emissions and hazardous and non-hazardous waste to zero.

e. **Human Rights:** Companies will regularly assess key risks related to human rights throughout their entire operations, and will employ management systems that are aligned with internal policies and support the implementation of universal standards.

8.5. SUPPLY CHAIN

Companies will require their suppliers to meet the same environmental and social standards as the company has established for itself. Companies will establish sustainable procurement criteria, catalyze improved supplier performance, and facilitate disclosure of suppliers' sustainability information.

1. **Policies and Codes:** Companies will set supply chain policies and codes aligned with overall social and environmental standards.
2. **Align Procurement Practices:** Companies will address sustainability performance in procurement criteria and contracting.
3. **Engaging Suppliers:** Companies will ensure that at least 75% of the company's Tier 1 and Tier 2 suppliers and 50% of Tier 3 suppliers meet the company's standards for sustainability performance.
4. **Measurement and Disclosure:** Companies will disclose a list of their Tier 1 and 2 suppliers and measure and disclose suppliers' sustainability performance.

8.6. TRANSPORT AND LOGISTICS

Companies will systematically minimize their sustainability impact by enhancing the resiliency of their logistics. Companies will prioritize low impact transportation systems and modes, and address business travel and commuting.

1. **Transportation Management:** Companies will develop transportation criteria that incorporate distance requirements from site to market and establish decentralized and localized distribution networks.
2. **Transportation Modes:** Companies will review logistics to prioritize low-impact transportation modes.
3. **Business Travel and Commuting:** Companies will decrease greenhouse gas emissions from business travel and employee commuting by 50% from a baseline of 2005.

8.7. PRODUCTS AND SERVICES

Companies will design and deliver products and services that are aligned with sustainability goals by innovating business models, allocating R&D spend, designing for sustainability, communicating the impacts of products and services, reviewing marketing practices and advancing strategic collaborations.

1. **Business Model Innovation:** Companies will innovate business models to reduce material inputs and prioritize a transition to sustainable products and services.
2. **R&D and Capital Investment:** Companies will use sustainability as a primary filter through which all R&D and capital investments are made. 50% of the R&D investment will be focused on developing sustainability solutions.
3. **Design for Sustainability:** Companies will approach all product development and product management decisions with full consideration of the social and environmental impacts of the product throughout its lifecycle.
4. **Marketing Practices:** Companies will align their marketing practices and product revenue targets with their sustainability goals, and will market their designed-for-sustainability products and services with at least the same effort as their marketing of other products.

5. **Strategic Collaborations:** Companies will collaborate within and across sectors to innovate and scale sustainable products and services, and contribute to the development of open source solutions.

8.8. EMPLOYEES

Companies will make sustainability considerations a core part of recruitment, compensation, and training, and will encourage sustainable lifestyle choices.

1. **Recruitment and Retention:** Companies will incorporate sustainability criteria into recruitment protocols, employee performance processes, compensation, and incentives.
2. **Training and Support:** Companies will develop and implement formal training on key sustainability issues for all executives and employees, and facilitate coaching, mentoring, and networks for sustainability knowledge sharing.
3. **Promoting Sustainable Lifestyles:** Companies will promote sustainable lifestyle choices across their community of employees through education and innovative employee benefits options.

CHAPTER 9

CO_2 MARKETS

CONTENTS

9.1. WHY CARBON MARKETS?

The 1980s and 1990s saw a developing familiarity with environmental change risks and the related need to decrease ozone depleting substance discharges. Starting with the main World Climate Conference in 1979, consideration finished with the marking of the UN Framework Convention on Climate Change (UNFCCC) at the 1992 Earth Summit in Rio de Janeiro. In that arrangement, 166 (presently 194) countries recognized the need to restrict the collection of ozone harming substances (GHGs) in the air to a level that will "forestall perilous anthropogenic obstruction with the environment system."1 During that equivalent period, discharges exchanging started arising as a viable and progressively famous approach device to address contamination control, especially with the fruitful phasedown of lead in fuel and making of the corrosive downpour exchanging program the United States (Tietenberg, 1985; Stavins, 1998). On the global scene, the Montreal Protocol on Substances that Deplete the Ozone Layer, endorsed in 1987, laid clear preparation for targets and plans for discharges levels in various nations, and it incorporated a restricted measure of outflows exchanging. It is subsequently not unexpected that there was impressive excitement for utilizing this apparatus to address environmental change as nations wrestled with how to plan the 1992 UNFCCC and afterward meet its goals during the 1990s. To be sure, support for worldwide GHG discharges exchanging started in the last part of the 1980s and mid 1990s: the United States at first advanced it in the UNFCCC deal talks, and "joint execution" as a casual adaptation of emanations exchanging eventually showed up in the UNFCCC (Wiener, 2001).

This excitement for emanations exchanging eventually defeated different protests and prompted the marking of the Kyoto Protocol in 1997. The Kyoto Protocol was the first vehicle for discharges exchanging quite a while—for sure we will call carbon markets. Kyoto set up an arrangement of outflows limits for a container of six GHGs for created nations, components for those created nations to exchange their outflows cutoff points, and systems for created nations to balance their discharges by funding discharges decreases in agricultural nations. While the Kyoto Protocol itself has prompted a tiny number of exchanges straightforwardly among nations, the European Union (EU) and an assortment of different wards have since sought after discharges exchanging to decrease their territorial GHG outflows. Carbon markets are currently the biggest class of natural or discharges exchanging markets the world as far as both volume and market esteem, by an extremely significant space. However, how viable has emanations exchanging been attending to environmental change? What are the distinctive elements of exchanging discharges of ozone depleting substances contrasted with, for instance, regular air or water contamination? What have we realized as carbon markets have been planned, carried out, and worked? The reason for this paper is to respond to these inquiries and feature new arising issues that presently should be stood up to. Following eight years of carbon market experience following the production of the EU Emissions Trading System in 2005 experience that incorporates developing business sector volumes, market worth, and emanations inclusion—some portion of that answer should be, indeed, carbon markets can work successfully. Notwithstanding, we have gained tons of useful knowledge, and there are significant illustrations for flow and future policymakers, experts, and specialists. The following area gives an outline of the standardizing hypothesis of carbon market configuration, starting with a portrayal of a portion of the applicable traits of GHGs. Area three talks about genuine encounter with existing and proposed discharges exchanging program plans, while segment four portrays and attracts illustrations from functional experience those business sectors. Area five looks forward and portrays central questions confronting emanations exchanging strategies.

9.2. NORMATIVE THEORY OF POLICY DESIGN FOR CARBON MARKETS

There are various distinctive characteristics of the environment issue that are applicable to the plan of carbon markets and that separate them from most other outflow or asset markets. The first is the worldwide idea of the environment issue. GHGs are one of a handful of the instances of a worldwide, consistently blended poison: GHG emanations all through the world have similar results paying little heed to where they are emitted. Put another way, GHG outflows have the very same externality properties across nations that numerous customary discharges have inside a neighborhood ward. It is inadequate to control the GHG discharges from a specific source or locale to diminish the dangers to that district. This component of the environment issue requires that, at last, a globally planned approach should be taken, with suggestions for carbon advertisements that make them strange in the event that not remarkable among natural business sectors. Second, GHGs are seemingly perpetual stock contaminations, staying in the climate on the request for a long time to hundreds of years. It is the amassed worldwide environmental grouping of GHGs, as opposed to a sum discharged at a specific moment, that is connected to a dangerous atmospheric deviation and climatic change. Similarly, the capital stock that produces, disseminates, and consumes GHGs additionally will in general be enduring; vehicles and significant machines can endure north of 10 years, while power plants and structures most recent a very long while. It is consequently critical to remember a drawn out worldwide viewpoint while tending to the environment issue, in any event, while thinking about specific close term provincial strategies. Third, while carbon dioxide is the major GHG, there are various GHGs of fluctuating strength (measured as far as how much intensity energy caught by a given measure of each gas) and life span, from methane, with a long period of 12 years, to sulfur hexafluoride, with a long period of 3,200 years. Fourth, GHGs are unavoidable in the economy, as opposed to being related to a specific arrangement of sources, areas, or advances. Carbon dioxide, the predominant GHG, is a major result of the burning of petroleum derivatives (coal, oil, and flammable gas) for energy creation. Energy is utilized all over, and petroleum derivatives are the wellspring of more than 80% of U.S. also, worldwide energy utilization. As a result, the potential market size is a lot bigger than other existing ecological business sectors. For instance, in 2008, petroleum product ignition represented around 30 billion metric lots of worldwide carbon dioxide emanations (Boden et al., 2011). The market worth of one year of recompenses for these discharges at $10 per metric ton would be $300 billion; at $25 per metric ton it would be $750 billion. At higher remittance costs, or when collected across a few vintages of stipends, the worth is effectively in the trillions of dollars. As another model, the assessed market worth of remittances that would have been made, simply in the United States, by proposed Senate regulation in the 110th Congress was $6–7 trillion (Samuelsohn, 2008). A connected point is that while adjustment of GHG focuses in the environment at last suggests driving net ozone harming substance discharges to approach zero, the not so distant future probably involves huge proceeded with emanations. The inescapable financial job of fossil energy utilization combined with restricted other options and proceeded with worldwide monetary development implies that GHG discharges won't be decreased similarly that ozone-draining substances were essentially killed during the 1980s and 1990s (Sunstein, 2007). In situations with restricted to direct discharges decreases, complete alleviation costs will quite often be a little part of the recompense esteem (Burtraw and Evans, 2009). This prompts unexpected distributional outcomes in comparison to additional commonplace circumstances where moderation is a huge portion of complete outflows. A fifth significant component is that choices exist for counterbalancing GHG discharges, in this way invalidating or possibly essentially diminishing their effect on the environment. Counterbalances, taking everything into account, incorporate any supported strategies for decreasing outflows or eliminating GHGs straightforwardly from the climate (e.g., through ranger service) that are not in any case covered under a specific cap. On a fundamental level, balances can happen inside or beyond the purview covered by an emanations cap. On a connected note, carbon catch and capacity (CCS) advances offer a finish of-pipe arrangement, keeping away from climatic discharges by catching carbon dioxide from cycle or burning gases and putting away it underground. While end-of-pipe approaches are very normal in customary air and water contamination control, CCS is viewed as a somewhat progressed approach in the environment setting and is still in the advancement stage. Regardless, just like with counterbalances, well designed strategy would require a component for crediting CCS. Given the level of intricacy and long time spans intrinsic in the environment issue, vulnerability along a few aspects—including the adequacy of various methodologies, the expense of alleviation, the advancement of innovation, and the environment takes a chance with themselves—will in general be an overall and unavoidable part of strategy examination, development, and execution. At the point when put with regards to the monetary worldview, the above ascribes of the environment issue have various ramifications for carbon market plan. We talk about the most significant of these in this part, including the significance of consistency adaptability for cost-adequacy; the meaning of advantage and cost highlights for proficient instrument plan; and the connection between stipend portion, government income and use, distributional effects, and worldwide rivalry. For additional detail on a few of these plan issues, Aldy et al. (2010) give an exhaustive survey of the writing on planning environment moderation strategy.

9.3. COST-EFFECTIVENESS, COMPREHENSIVENESS, AND FLEXIBILITY

Cost-viability, or accomplishing a given total GHG moderation focus at the most minimal conceivable expense, will in general be a top-level worry for market-based strategy plan. To hypothetically empower cost-viability (albeit not observationally promise it), the above qualities will quite often highlight the plan of carbon showcases that have complete inclusion (i.e., across all gases, areas, sources, and advancements), can make the most of alleviation amazing open doors at a worldwide scale, and have a drawn-out time skyline and construction that supports coordinated decision making across time. This is once in a while alluded to as "what, where, and when adaptability," which supports a standardizing inclination toward exhaustiveness and insignificant hindrances to exchanging open doors across space, time, and moderation movement. "What adaptability" connects with the breadth of any carbon market, with hypothesis directing plan toward consideration of as many sorts of GHGs (e.g., carbon dioxide, methane, nitrous oxide, and fluorinated compounds), areas (e.g., power, transport, industry, and farming), and advancements (e.g., fuel exchanging, CCS, and ranger service and other biosequestration) as is attainable. One admonition is that it is critical to think about what occurs beyond a worldwide exchanging framework when just a few purviews are chasing after outflows exchanging. Specifically, "spillage" can happen when guideline inside one locale prompts expanded imports from another, unregulated purview—a peculiarity that might apply to an assortment of ecological approaches, not just carbon markets. Spillage will in general be packed in energy-concentrated, trade sensitive areas and has the two discharges and monetary outcomes as financial action shifts across locales. There are an assortment of ways of tending to these worries (Fischer and Fox, 2009). "Where adaptability" addresses the huge variety in the expenses of GHG relief across districts and nations. Whenever zeroed in on planning a carbon market for a particular locale, it is subsequently essential to consider the degree to which stipends as well as counterbalances from different districts will be allowed to be utilized for consistency. Insofar as the legitimacy of such recompenses can be laid out, cost-adequacy would will quite often highlight open exchange remittances across purviews. Notwithstanding, this sort of linkage can confront down to earth and political obstacles, to a limited extent because of the related asset moves, as we will talk about beneath (see Section 5.1). At long last, given the stock poison nature of GHGs, outflows on a specific day, month, or year are not important for environment influence. Therefore, permitting "when adaptability" through banking or getting of stipends across time will in general inflate cost-adequacy without extra damage to the environment. Inasmuch as the full scope of alleviation choices across reality is accessible to the market framework, exchanging will chief lead to a financially savvy allotment of contamination control activities through the end of chances for exchange.

9.4. EFFICIENCY AND INSTRUMENT DESIGN

Notwithstanding cost-viability, which suggests evening out of minimal contamination control costs, by, and large monetary productivity requires adjusting the negligible advantages and peripheral expenses of GHG alleviation. One of the upsides of a market-based cap-and-exchange way to deal with contamination control is that—for a given outflows target—it requires no ex-bet information on contamination control expenses or advantages. Nonetheless, on the off chance that there is interest in adjusting the peripheral expense and advantages of the arrangement, monetary data is expected for the two sides of this record. This raises unexpected issues connected with assessing the expenses of relief, the adapted worth of environment harms (Interagency, 2009; National Research Council, 2010), and the plan of strategy instruments that are effective notwithstanding vulnerability over expenses and advantages. The early monetary writing on environment strategy instrument decision under vulnerability (Pizer, 2002; Newell and Pizer, 2003) highlighted the upside of cost-based (i.e., a carbon charge) over amount based instruments (i.e., cap-and-exchange) in light of a changed Weitzman-type contention (Weitzman, 1974) and accepting consistency must be accomplished on a yearly premise. This prior work likewise showed that cost like changes inside a cap-and-exchange program—roofs and floors on the recompense cost or in any case changing the cap to oblige cost shocks—could accomplish a similar result as a carbon charge (Newell et al., 2005; Murray et al., 2009). Additionally, late work has shown that conventional financial arrangements—which are remembered for practically all emanations exchanging programs—may come near the proficiency of a carbon charge minus any additional changes (Fell et al., 2012). One of the prickly issues raised by the likely development of remittances across time is deciding the exchanging proportion to be applied to banked or acquired stipends and choosing how to apply it. Hypothesis proposes that the ideal exchanging proportion between periods is equivalent to one or more the rebate rate short the ideal pace of progress in license costs (Leiby and Rubin, 2001). Notwithstanding this recipe, a rebate rate is required, which raises a particular arrangement of scientific difficulties for both the assessment of harms and the rate at which the carbon cost ought to rise (Interagency, 2009; Aldy et al., 2010; National Research Council, 2010). Because of the drawn out nature of the environment issue, impetuses for mechanical development in GHG alleviation innovations are likewise significant for productivity (Newell, 2010).

As for carbon market plan, a central point of interest for innovation advancement and arrangement is setting carbon market toughness at a level that creates the ideal motivator through the cost system. Within the sight of advancement overflows, one may be enticed to support rigidity past the level directed essentially by the environment externality. Notwithstanding, an all the more financially productive methodology is to perceive the requirement for more than one approach instrument to address these various market blemishes and incorporate direct development strategy (Jaffe et al., 2003). Instrument plan with regards to various market flaws and different strategy instruments can become intricate, as the ideal degree of toughness of a specific instrument relies upon the levels of the others (Fischer and Newell, 2008).

9.5. ALLOWANCE ALLOCATION: DISTRIBUTIONAL IMPACTS AND EFFICIENCY

Similarly as with other market-based programs, GHG stipends can be sold, designated free of charge, or planned with a mix of the two. Remittance assignment has distributional and effectiveness results, and these can be enormous—on the request for many billions of dollars—given the sizable monetary rents in question. As indicated above, due to the generally little proportion of alleviation to outflows in most carbon exchanging guidelines to date, installment of these rents will in general be the predominant distributional expense. In many key carbon-transmitting areas, we would anticipate that serious strain should lead item costs to reflect carbon content, no matter what any free allocation. Consequently, end clients of energy would eventually wind up paying these rents. Thus, there can be critical distributional effects on elective free designation approaches and equations that appropriate these rents, with suggestions for attainability and value across firms, pay gatherings, districts, and ages. Except if utility or other guideline intercedes, free recompense portions will generally accumulate to firms and result in higher value values, to the advantage of higher pay gatherings and to the detriment of lower pay gatherings (Dinan and Rogers, 2002). Indeed, even without the potential value effect of free allotment, it is genuinely deeply grounded that carbon evaluating (through either an unloaded remittance market or expense) excessively hurts low-pay families in created nations (Hassett et al., 2009). While higher-pay families will quite often burn through more on effort than do low-pay families, they spend less as a negligible part of pay. Various systems have been proposed to improve the relapse of carbon valuing, including singular amount discounts (e.g., supposed cap-and-profit), assignments to power merchants, and equal counterbalancing changes to pay or federal retirement aide charges (Burtraw et al., 2010). According to a productivity perspective, better remuneration instruments will quite often try not to bring down energy costs (to protect suitable impetuses) and make the most of chances to bring down other distortionary charges. Missing a worldwide exchanging program, free remittance portions for energy-escalated businesses are in many cases a vital component to address political achievability as well as to keep away from emanations spillage for negligible makers. For moderate arrangements, some examination on this issue has found that, in the United States, something like 15–20% of free stipends are expected to remunerate energy escalated businesses for their deficiency of maker excess (Bovenberg and Goulder, 2001). In any case, other exploration has shown that there is critical variety in consistency costs inside energy escalated areas—making the two victors and failures under a carbon market—so it can require a few times higher part of remittances to repay the washouts than would be demonstrated by the normal misfortune in maker excess (Burtraw and Palmer, 2008). Productivity issues emerge mainly concerning how much income is raised through sales and the amount of this income is reused to counterbalance other distortionary burdens or utilized for different purposes (e.g., innovation projects, transformation, or energy effectiveness programs). The writing will in general help the view that income nonpartisan full unloading and the utilization of continues to bring down other distortionary charges both deal a huge expected gain in effectiveness comparative with free distribution or single amount reallocation of incomes (Aldy et al., 2010). There are quite a few alternate ways that carbon recompense sell off incomes could be utilized, and the expected requirement for assets to help correlative GHG relief endeavors—, for example, subsidizing energy effectiveness projects and innovation exploration, improvement, and exhibit—has connected these projects to carbon market plan. It isn't clear there is any productivity benefit to this linkage, however according to a commonsense perspective, the utilization of carbon market income to help other GHG alleviation programs has had appeal to policymakers and has appeared as a conspicuous component of both proposed and executed carbon market plan. Notwithstanding, connections between carbon markets and different strategies pointed toward lessening emanations may—in specific situations—bring about diminished discharge decreases, discounted cost-effectiveness, as well as an abatement in the market's recompense costs (Goulder and Stavins, 2012).

9.6. PROGRAM OVERVIEW AND KEY DESIGN CHOICES

The regularizing hypothesis talked about above centers around cost-adequacy, controllers fostering each program have needed to wrestle with huge partner interest in many plan decisions. The most hostile issues have spun around which areas to incorporate; how to limit serious misfortunes to firms; how to restrict purchaser cost builds; whether and how to utilize outflows balances; and the degree to and way in which firms exchange remittances. This segment investigates those key inquiries, summing up the underlying plan decisions in an assortment of discharges exchanging projects and featuring prominent plan components. Table 1 gives a wide outline of focal plan components in an assortment of past, present, and (for the United States) proposed carbon exchanging frameworks. Moreover, this segment addresses the institutional and political accounts of significant projects. Despite the fact that we can't do full equity to the wide cluster of political and monetary decisions prompting every choice, we note a few key factors that impacted starting plan decisions. While all carbon exchanging programs are moderately youthful, some have changed significantly since sending off. We depict those significant changes, and the market encounters that provoked them, in area four as we talk about market experience all the more by and large.

9.6.1. Program Overview

9.6.1.1. Kyoto Protocol

The Kyoto Protocol, embraced in 1997, laid out the primary non-deliberate carbon market, committing specific countries to meet GHG discharges decrease targets and laying out a structure for remittance exchanging across global lines. The convention went into force in February 2005. 37 industrialized signatories, known as Annex-I countries, are answerable for diminishing emanations by determined targets (United Nations Framework Convention on Climate Change 2012b). On the off chance that a country can't meet its objective, it might either buy stipends (called Assigned Amount Units, or AAUs) from an individual Annex-I country or buy emanations counterbalances from projects that decline discharges in different regions of the planet.

The Kyoto Protocol laid out two significant instruments to balance outflows: the Clean Development Mechanism (CDM) and Joint

Implementation (JI). CDM projects, which decrease emanations in non-industrial countries, procure one Certified Emissions Reduction credit (CER) per metric ton of GHG discharges decrease, which might be bought by countries or firms to meet their commitments under the convention. JI projects additionally acquire one credit for each metric ton of outflows decrease (called Earned Reduction Units, or ERUs) and come basically from projects in the previous Soviet Union. Both venture types look to energize clean energy speculation and learning while at the same time permitting Annex-I countries adaptability in gathering their emanations targets (United Nations Framework Convention on Climate Change, 2006). While these offset programs keep on working, the fate of the Kyoto Protocol as a structure for diminishing outflows is questionable. Discussions in Durban, South Africa, in late 2011 finished with countries consenting to concur by 2015 on a far reaching intend to lessen emanations that would produce results by 2020 under the more extensive UNFCCC. In 2012, talks in Doha, Qatar brought about the augmentation of the Kyoto Protocol from 2013–2020. Notwithstanding, the United States never approved the convention; China (the world's biggest producer) isn't expected to decrease emanations under it; and Canada, Japan, and Russia have not consented to take on extra responsibilities under Kyoto after 2012 (United Nations Framework Convention on Climate Change 2012b). Australia has focused on unobtrusive emanations decreases by 2020, promising to establish further cuts in the event that the world's other significant producers focus on an "aggressive worldwide arrangement" (United Nations Framework Convention on Climate Change, 2011). New Zealand has expressed that it won't commit to any responsibilities under the second round of the Kyoto Protocol, however it stays resolved to its general discharge decrease objectives. These changing responsibilities paint a questionable picture pushing ahead. Regardless, an assortment of states have created market-based projects to decrease their GHG outflows inside or beyond the Kyoto structure. Why these states have decided to act missing areas of strength for a system is an inquiry we don't zero in on; rather perusers ought to see work by Aldy et al. (2007) on global environment designs, and Victor (2008) and Nordhaus (Nordhaus and Boyer, 1998) on issues with the Kyoto approach. The accompanying area gives an audit of significant projects.

9.6.1.2. EU Emissions Trading System

Albeit most European countries are inclined toward a carbon charge going into the Kyoto talks, the EU has made by a wide margin the world's biggest market-based framework to decrease GHG emanations: the EU Emissions Trading System (EU ETS). Truth be told, the EU ETS started working in 2005, three years before Kyoto's most memorable responsibility period started. The program has worked in stages, with a pilot stage from 2005 to 2007 covering the power area and certain weighty industry, a second stage from 2008 to 2012 growing inclusion somewhat, and a third stage set for 2013–2020 that adds a huge scope of modern movement alongside a few new gases (Ellerman et al., 2010; European Commission 2012a). Under the initial two stages, every one of 27 EU countries (later extended to 30) submitted National Allocation Plans (NAPs) to the EC. 9 Each NAP set an outflows focus for covered areas and was explored by the Commission, which had the ability to change the quantity of stipends apportioned to every EU part. The program's underlying plan was profoundly decentralized, and when the NAPs were concluded, countries had critical circumspection on the most proficient method to convey outflows credits to various areas of their economies. The pilot stage was something of a test during which states and the private area could foster a superior comprehension of how the exchanging project would work as the cap fixed in ongoing stages. Rests were created under close cutoff times, and the Commission's acclimations to every public proposition were hampered by deficiencies of time and information. Regardless of the trouble of working out the same old thing (BAU) projections with fragmented information, the Commission created targets in light of an unobtrusive discharges decrease objective (Ellerman et al., 2010). Indeed, even with unassuming decrease targets, part countries stressed that evaluating carbon could hurt modern areas subject to worldwide rivalry and raise shopper costs for homegrown power. Therefore, by far most of European Union Allowances (EUAs) were designated for nothing in the pilot and the subsequent stages. Every country, through its NAP, resolved the level and conveyance of free distribution to various areas through 2012. Rests likewise indicated the quantity of balanced credits producers in every country could buy from CDM or JI projects, with limits going from 0 to 20% (European Commission, 2009).

Market insight and examination and investigation in the EU ETS have prompted huge changes as the program enters its third stage in 2013. We talk about that experience and changes in remittance designation, offset rules, and the capacity to bank credits in area four.

9.6.1.3. United States

The United States, which marked yet never confirmed the Kyoto Protocol, has seen many authoritative endeavors to address GHG discharges. Albeit the viability of an emanations exchanging approach was first exhibited for a huge scope by the United States' sulfur dioxide and nitrous oxide programs under the 1990 Clean Air Act Amendments, the U.S. Congress has not passed a GHG exchanging program (Burtraw et al., 1998; Meckling, 2011). Expanding on an assortment of past endeavors starting as soon as 2003's Climate Stewardship Act (co-sponsored by Senators Joseph Lieberman and John McCain), the U.S. Place of Representatives barely passed the American Clean Energy and Security Act in 2009 (also called the Waxman-Markey bill, after the bill's cosponsors). The Waxman-Markey bill would have laid out an economy-wide outflows exchanging program, which zeroed in at first on the power and modern areas, extending following quite a while to incorporate transportation, and permitting producers to counterbalance up to 30% of their covers from either U.S.- based or worldwide offset projects (Larsen et al., 2009). In any case, political resistance bound the regulation in the Senate, where it was rarely taken up (Meckling, 2011). In spite of the shortfall of a cross country framework, a few districts had previously set up or created plans to carry out their own GHG exchanging programs. In 2005, seven northeastern states turned into the principal assortment of purviews in the United States to consent to an outflows exchanging program: Connecticut, Delaware, Maine, New Hampshire, New Jersey, New York, and Vermont. Maryland joined in 2006, Massachusetts, and Rhode Island joined in 2007, and New Jersey pulled out toward the finish of 2011. Known as the Regional Greenhouse Gas Initiative (RGGI, articulated as "Reggie"), this program just covers huge power generators, and almost 100% of stipends are unloaded. Incomes from remittance barters go to state legislatures, which are expected to contribute somewhere around 25% of those assets toward energy proficiency or environmentally friendly power programs. Counterbalances for producers are restricted to simply 3.3% and come from projects for the most part inside RGGI states, however no offset projects have yet been utilized for consistency (Regional Greenhouse Gas Initiative Inc. 2012). A second U.S.- based emanations exchanging framework started in California in 2013. Arizona, California, New Mexico, Oregon, and Washington started the Western Climate Initiative (WCI) in 2007, later adding Utah and Montana along with British Columbia, Ontario, Manitoba, and Quebec (we talk about these Canadian areas beneath). Be that as it may, no WCI states in the U.S. other than California have carried out a discharges exchanging program

(Craig, 2011). The WCI has created a definite strategy proposition to arrange future projects executed by its individuals, with an eye toward connecting state programs under one umbrella. California's discharges exchanging program, ordered by regulation, utilizes arrangements created by the state's Air Resources Board in meeting with the WCI and will sell around 66% of its recompenses in 2013. California's program will at first cover the power area and huge modern sources, representing approximately 37% of the state's GHG outflows, and extend to cover 85% by 2015, including transportation fills (California Code of Regulations, 2011). A last local understanding, the Midwestern Greenhouse Gas Reduction Accord (MGGRA), was endorsed in 2007 by six lead representatives. The agreement, which included Illinois, Iowa, Kansas, Michigan, Minnesota, Wisconsin, and Manitoba alongside a few eyewitness states, was to work like the WCI, assisting its individuals with laying out programs by 2012. In 2010, MGGRA delivered an itemized model rule for part states to adjust and utilize. The proposition covered the power area, major modern producers, and transportation fuel merchants, with most stipends being assigned for nothing at the prudence of part states (Midwest Greenhouse Gas Reduction Accord, 2010). Regardless of the MGGRA's advancement, new political initiative and moving monetary needs finished the work in mid 2011 (Volvovici, 2011)

9.6.1.4. Canada

Canada approved the Kyoto Protocol in 2002 however never passed through regulation to direct outflows. Driven areas of strength for by development and a roaring regular assets area, CO_2 emanations surpassed their objectives, and Canada pulled out from the Kyoto Protocol in 2011 (Canadian Government 2011b). Numerous Canadian policymakers are hesitant to follow up on environmental change without the United States making a practically identical move. Regardless, a few areas have put forth autonomous attempts to lessen GHG discharges (Canadian Government 2011a; De Souza, 2011). Quebec is at present on target to carry out a GHG emanations exchanging program beginning in 2013. Since practically all of its power comes from non-carbon-transmitting hydropower, the program won't influence most power generators however will apply to modern sources (Center for Energy, 2012). Quebec fostered the framework utilizing arrangements created under the WCI, permitting it to connect with California (for a more nitty-gritty conversation on connecting, see area 0). English Columbia (which executed a carbon charge in 2008), Manitoba, and Ontario, additionally WCI individuals, plan to embrace exchanging programs the future however have not indicated dates (Western Climate Initiative 2012). Manitoba was additionally an individual from the now-ancient MGGRA. The asset rich region of Alberta, which produces more than 40% of Canada's GHG outflows, has taken on a methodology in view of decreasing energy force. Huge producers, including oil makers, should lessen their discharges per unit of result by 12% against a differing gauge estimation somewhere in the range of 2007 and 2010. In the event that they can't meet these objectives, producers can consent by one of three components: buy overabundance credits from another producer, buy Alberta-based counterbalances, or pay C$15 per metric ton toward an energy/environment innovation reserve. Most stipends are distributed openly to producers, and, starting around 2010, the program covered 97 offices that represented approximately 50% of Alberta's GHG outflows (Alberta Government 2011a, b, 2012). The program's unassuming power decrease targets implies that regardless of whether producers meet their objectives, generally speaking, GHG outflows might well increment (for a basic conversation of this issue, see Doluweera et al. (2011)).

9.6.1.5. New Zealand and Australia

New Zealand sent off an emanations exchanging program 2008, which was intended to cover practically all New Zealand outflows with covers in light of its 2008–2012 responsibility under the Kyoto Protocol. Since New Zealand is a little economy, the program was worked around connecting to different business sectors; this at first incorporates the Clean Development Mechanism (CDM) yet could be extended to other public or local carbon markets (like the EU or Australia). This component has made the program helpless against global strategy vulnerability and to issues encompassing the CDM. The program covers a somewhat modest number of huge producers who should decrease discharges, buy homegrown or global counterbalances, or pay $25 (New Zealand) per ton of emanations. The program has no value floor and costs have consistently declined through 2012, by and large following the developments of CDM costs. Enterprises confronting worldwide contest, agriculture, and fishing get up to 90% free assignment, however the power area, transportation, and ranger service don't (New Zealand Government 2012a). In Australia, after a long and quarrelsome political cycle where the state head switched her public position, a progression of bills became regulation in 2011 that will start an outflows exchanging program 2015. Meanwhile, significant carbon producers will pay a consistently expanding carbon charge set by the regulation. Government incomes from this expense and sales starting in 2015 will go to new spending on proficiency, renewables, and innovation, with something like 50% of incomes going to expanded benefits installments, expanded tax breaks, and diminished annual charges for families (Australian Government 2012b). In August 2012, Australia reported that its program will connect with the EU ETS, permitting producers to give up EUAs to consent to up to 50% of their necessities (Australian Government 2012a; Reklev 2012a). Furthermore, European producers might involve Australian stipends for consistence as soon as 2018. In any case, Australia's ongoing resistance has made canceling the carbon value "the first concern" on its plan, raising doubt about the approach's feasibility pushing ahead (Australia Liberal Party, 2012).

9.6.1.6. Other Markets, Including Voluntary Markets

Ongoing regulation passed in South Korea and Mexico has laid the basis for new public level projects starting in 2015. Much about these projects still needs to be chosen, however one intriguing element of Korea's program is that it won't permit worldwide balances for consistence until somewhere around 2020 (Reklev 2012b). Furthermore, some have guessed that Mexico's new president, Enrique Peña Nieto, is probably not going to carry out a significant part of the as of late passed environment regulation (Teixeira, 2012). China has laid out a progression of nearby and provincial experimental runs programs, while programs being talked about in India, Japan, Vietnam, and Thailand show an interest in cap and exchange across quite a bit of Asia. Different emanations exchanging recommendations are as of now being talked about or advancement Brazil and Chile, among others (Hood, 2010). At last, deliberate carbon markets allude to an assortment of associations that permit people or organizations to buy balances from discharges decrease projects. Beginning around 2002, intentional business sectors have developed from $43 million in incomes to a pinnacle of $705 million out of, 2008; they remained at $572 million starting around 2011 (Ecosystem Marketplace and Bloomberg New Energy Finance 2008–2012). Many associations offer deliberate carbon counterbalances, and their guidelines for assessing and observing GHG decrease projects are ordinarily less rigid than those utilized for the CDM or JI. One advantage of less severe guidelines is diminished administration and the potential for lower project costs; notwithstanding, more fragile principles could likewise prompt confirmation of ventures that don't give their expressed advantages (Benessaiah, 2012).

9.6.2. Key Design Choices

9.6.2.1. Setting the Cap and Comparability Issues

A central inquiry for any cap-and-exchange program is the level of the cap. Locales have adopted an assortment of strategies to setting decrease targets and estimating progress. As verified in Table 1, a few projects expressly base cap levels on the Kyoto Protocol targets and 1990 base year, yet many have created elective targets and base years. Missing a typical way to deal with setting targets, sharing liability regarding outflows decreases across wards, or estimating progress, it is hard to contrast the objectives of one program with the following without debate. Adding to the equivalence challenge is the way that most projects don't cover all financial action. For instance, a 21% decrease in the cap determined by the EU ETS doesn't address a 21% GHG discharges decrease across the whole economy. All things being equal, the EU ETS is one of a few apparatuses executed by the EU to arrive at its Kyoto Protocol targets. Australia and New Zealand, then again, cover all or practically all monetary action with their exchanging frameworks and will endeavor to arrive at their vows under the 2011 Copenhagen Accord primarily through emanations exchanging. Most sub-public projects in the United States and Canada neither cover the whole economy (with the eminent exemption of California, which will do as such) nor unequivocally interface their decrease objectives to any peaceful accord. Obviously, specialists face an assortment of difficulties in looking at provincial, public, and subnational programs, particularly when those projects are intended to resolve a worldwide issue.

9.6.2.2. Regulated Sectors and Allocation

Since greenhouse gas emissions occur as a by-product of virtually every form of economic movement, states face the urgent inquiries of which areas to manage and whether to sell stipends or distribute them for nothing. After a mid-inclination to give free designations, new projects and the reexamined EU ETS have advanced toward unloading recompenses (we talk about certain explanations behind this in area 0). The EU ETS, RGGI, and different projects have zeroed in on covering enormous point sources, for example, power plants and industrial facilities. Such sources have generally low observing expenses and—according to a political viewpoint—are frequently simpler to focus for guideline than sources like transportation. Also, most electric power markets don't confront significant rivalry from unfamiliar firms, implying that many power makers can pass along expenses for buyers. As a result, power generators in many projects get less free recompenses than different sorts of producers. The NZ ETS, for instance, gives no free recompenses to the power area, and RGGI has given only 4% in its 2009–2011 period. Weighty modern cycles (e.g., concrete, aluminum, lime, and fuel purifiers) are covered by each framework other than RGGI, yet, dreading monetary separation, legislatures have given significant free distributions to these organizations. In each exchanging program that incorporates them, these commonly exchange touchy modern areas get up to 95% free distribution. Controllers by and large compute an area or company's degree of free portion in view of equations including openness to unfamiliar rivalry and every area's capacity to decrease their outflows force. It is actually important that a significant portion of these free stipends might be going to makers who don't need free allotment to work locally. Therefore, these makers procure "bonus benefits."

9.6.2.3. Cost and Volatility Containment

We start the conversation of expenses by taking note of that market-based emanations decrease programs ordinarily offer significant investment funds over conventional principles based guidelines (Stavins, 1998; Carlson et al., 2000). Among market-based programs, regularizing hypothesis underlines the worth of one or the other fixing or compelling costs, as indicated in area 0. However, the political discussion and press inclusion has placed more spotlight on keeping away from excessive costs. This might be obvious, considering that a market-based program unavoidably communicates costs significantly more straightforwardly than standards based guideline. No matter what the inspiration, huge vulnerability over decrease costs has driven program originators to incorporate components to keep recompense costs from surpassing financially and politically decent levels. The discharges exchanging programs examined above have normally gone to at least one of the accompanying three sorts of cost control. To start with, controllers can force a cost roof, permitting producers to buy recompenses straightforwardly from the public authority at the maximum price tag. For instance, members in California or Quebec's program will actually want to buy credits for $40-$50/metric ton, basically covering exchanging costs (Western Climate Initiative, 2012). Second, producers might be permitted to bank or acquire remittances from past or future consistence periods. This choice permits producers more noteworthy adaptability in gathering discharges decrease focuses after some time without constraining them to exactly accomplish every break focus (as indicated in area 0, this sort of adaptability is legitimized on cost-viability grounds alone). The EU ETS, which started with no banking or getting between stages, presently permits limitless banking of credits (Ellerman et al., 2010). In spite of the fact that it would likewise diminish consistence costs, acquiring arrangements have been less broadly taken on (Fell and Morgenstern, 2009).

Third, high market costs can set off arrangements that loosen up the limitations of the program. In RGGI, for instance, assuming carbon costs reach \$7/metric ton, producers are permitted to buy more carbon balances to meet their consistency needs. On the off chance that costs reach \$10/metric ton, producers might buy even more counterbalances to arrive at their objectives (Regional Greenhouse Gas Initiative Inc. 2012). A few projects additionally utilize value floors to keep market costs from falling under a specific level. Closeout cost floors are utilized in RGGI and California alongside expected programs in Australia and Quebec. These floors are not executed as a stipend repurchase, however rather set a cost underneath which recompenses won't enter the market. Value floors might look to achieve two objectives: one, to keep up with costs at a level where firms have a motivator to put resources into emanations decrease innovation, or two, to give a consistent wellspring of income for states.

9.6.2.4. *Offsets*

Global emanations counterbalances offer an extremely huge likely pool of alleviation open doors, giving industrialized countries a cheaper choice for GHG decreases comparative with lessening discharges inside their own boundaries (Weyant and Hill, 1999). Homegrown or nearby balances can likewise offer expense reserve funds comparative with open doors inside a given cap-and-exchange program, yet address a more modest universe of exercises contrasted with global counterbalances. Albeit explicit arrangements and limitations change, all projects to date utilize balances in some limit. Counterbalances from the CDM and JI assume a significant part in the EU ETS, for certain countries permitting up to 20% of discharges decreases to be met with balances. (Ellerman et al., 2010). Members in the EU ETS may not buy CDM or JI balances from ranger service or land-use change projects, due generally to inquiries over the changelessness of carbon sequestration in ranger service projects (Kim et al., 2008). Moreover, some have raised worry over likely mischief to native people groups because of ranger service projects (Stickler et al., 2009). While global balances have assumed by a wide margin the biggest part to date, territorial or nearby offset projects to decrease discharges exist in a few laid out and arising exchanging programs. The thought that balances ought to happen inside a program's boundaries, rather than from global CDM or JI projects, has grabbed hold now and again. Balance projects based inside a program's boundaries guarantees that the related speculation remains nearby home, to the advantage of neighborhood economies. Then again, neighborhood offset undertakings might cost more than balances gave abroad, inferring greater expenses for privately controlled enterprises. In New Zealand's program, where no sales are held, producers follow any neglected decrease objectives by buying balances, either from New Zealand-based ranger service counterbalances or from CDM or JI projects (New Zealand Government 2012b; Point Carbon 2012b). Local projects in North America, like RGGI, California, and Quebec, have thus far stayed away from the CDM, rather giving inclination to balances from territorial or homegrown discharges decrease projects. These North American projects permit balances to make up a little portion of consistency, so firms don't have to draw on the large numbers of metric lots of counterbalances accessible from the CDM or JI. Maybe more significantly, CDM and JI credits have just been accessible to parties with responsibilities under the Kyoto Protocol.

9.6.2.5. Market Monitoring and Oversight

After the 2008 monetary emergency, basically all monetary business sectors went under new investigation. Carbon markets were no special case, and new proposition for exchanging programs the United States accompanied calls areas of strength for. Truth be told, the 2010 Dodd-Frank monetary change and customer security bill made an interagency working gathering to direct a concentrate on keeping up with and expanding straightforwardness for carbon markets (Interagency Working Group for the Study on Oversight of Carbon Markets, 2011). Essentially, an EU mandate took on in 2011 (the Markets in Financial Instruments Directive, or MiFID) will fundamentally extend oversight of carbon markets (European Commission 2012b). Essential objectives for market oversight incorporate working with cost disclosure, guaranteeing straightforwardness and admittance to data, and forestalling control or maltreatment in the commercial center. Observing and oversight happens in the essential (first buy or issuance), auxiliary, and subordinates markets. In essential business sectors, controllers normally try to follow starting responsibility for, the lead of stipend barters, and the creation and confirmation of balanced credits. For instance, RGGI tracks recompense proprietorship through its CO_2 Allowance Tracking System (COATS), utilizes an autonomous screen to audit barters, and requires part states to check emanations offset programs (Regional Greenhouse Gas Initiative, 2010). Oversight of optional business sectors, where spot exchanges happen and prospects contracts are made, presents an alternate arrangement of difficulties. Equivalent admittance to data is crucial to guarantee a fair commercial center, and projects look to guarantee this by reporting new market information at foreordained stretches. Furthermore, controllers might try to restrict the market force of any single substance by restricting situations, as found in California and Quebec's approaching projects. Subsidiaries in view of discharges recompenses are regularly dependent upon similar market oversight as different subordinate instruments. In the United States, oversight of these instruments and trades falls under the domain of the Commodity Futures Trading Commission (Interagency Working Group for the Study on Oversight of Carbon Markets, 2011).

9.6.2.6. *Market Experience and Lessons*

Having investigated the underlying plan highlights related with various carbon markets, we presently go to showcase experience and development. We center around the EU ETS, CDM, and RGGI, for which there is essentially quite a while of information yet we likewise remember ongoing data for New Zealand and California, where prospects contracts started exchanging 2011. We draw various illustrations in view of the experience to date, featuring cost levels, market activity, banking, portion, and incomes, spillage, and counterbalances. We don't zero in straightforwardly on the Kyoto Protocol, as its essential consistence instrument, called Assigned Amount Units (AAUs), are commonly exchanged one-off exchanges, arranged, and offered straightforwardly starting with one Annex-I country then onto the next (Aldrich and Koerner, 2012). Despite the fact that countries have exchanged large number of AAUs these exchanges, normalized trades have not arisen. All things considered, the Kyoto Protocol's essential commitment to outflows exchanging markets has been the advancement of the EU ETS and the CDM, which we truly do investigate. Carbon costs in all markets have been falling beginning around 2008 because of the ongoing monetary downturn. Regardless, volumes have been expanding, both concerning movement inside business sectors as well as the making of new business sectors. The EU ETS has overwhelmed the commercial center, with far more prominent volumes, liquidity, and cost unpredictability than some other market. The EU ETS is additionally the biggest source for CDM and JI credits. Furthermore, the EU framework is the one in particular where a huge optional market has created, with market members trading normalized agreements as long as five years ahead of time on an assortment of trades. While exchanging the EU ETS started generally with non-normalized over-the-counter (OTC) exchanges, trade based exchanging likely outperformed OTC volumes at some point in 2008, showing expanded degrees of normalization and liquidity (Ellerman et al., 2010).

1. **Lesson: Positive Prices Imply Emissions Abatement, But How Much Is Unclear**

   The simple presence of a reliably certain cost on carbon recommends that these exchanging programs are somely affecting way of behaving as to discharges levels. Program plan, fundamental fuel costs, and bigger monetary powers all affect conduct of market members, however due principally to the young people of the EU ETS and different projects, research on the degree of every one of these effects comparative with the carbon cost stays restricted. One method for moving toward the decrease question is to appraise outflows decreases in view of versatilities got from related examinations. An unpleasant examination of projections from the proposed U.S. Waxman-Markey regulation recommends outflows semi-flexibilities of 0.0015–0.0061 in the year 2015. That is, for each $10/metric ton expansion in the cost of U.S. CO_2 remittances, outflows during 2012–2015 would fall somewhere in the range of 1.5 and 6% contrasted and a situation with no cost on CO_2 emanations. Assuming comparative monetary elements are impacting everything in the EU ETS, a stipend cost of $16/metric ton (the Phase I normal for the EU ETS) would propose that the program brought about decreases of 2–9% contrasted and BAU. Existing examination conforms to this unpleasant computation. Observational examination on Phase I of the EU ETS recommends that during 2005–2007, emanations fell by 2–5% contrasted and BAU (Ellerman et al., 2010; Anderson and Di Maria, 2011). In the event that precise, this degree of decrease is reliable with the thought that Phase I of the EU ETS was less an aggressive carbon-decrease plan than an endeavor to acquire insight for future outflows decrease endeavors (to be sure, the Kyoto Protocol didn't produce results until, 2008). Since Phase II of the ETS has as of late closed, we need observational exploration on reduction for 2008–2012. Nonetheless, utilizing the semi-versatilities noted over, the $20/

metric ton normal cost of EUAs from 2008 through mid-year 2012 infers a decrease of 3–12% contrasted with a BAU situation without any outflows exchanging framework stage I or II. With costs tumbling to 33% of that level toward the start of 2013, the potential for outflow decreases in stage III still need to be worked out.

A vital inquiry—and at some point analysis—of current market-based strategies is how much they energize long haul interest in new innovations instead of exclusively transient fuel-exchanging and energy protection. Much has been expounded on the significance of long term venture to address environmental change and the possible viability of market-based approaches to drive it (Jaffe, 2002; Newell, 2010). In any case, markets for CO_2 might be excessively new to completely move the drawn out certainty to make those ventures, and early examination into the EU ETS recommends that such speculations might be restricted (Leiter et al., 2011). On account of RGGI, outflows have fallen well underneath the cap because of fuel exchanging and feeble financial development, which probably restricts interest in long haul emanation decreases.

2. **Lesson: Despite Some Rough Patches, Markets Have Generally Matured and Operated Effectively**

Given the young people of the carbon showcases, some learning and advancement is normal. Regardless of the exemptions noted beneath, business sectors have commonly worked well, empowering cost revelation, expanding market straightforwardness, and reflecting bigger financial patterns. Albeit more modest business sectors like RGGI (where the cap has been non-restricting) and New Zealand don't see high volumes of exchanging, market profundity and liquidity in the EU ETS is broad. One beginning test in the EU ETS was the accessibility and nature of pattern outflows information. The EU ETS's pilot stage covers were developed under time tension and with a lack of dependable information. Whenever the principal tranche of genuine outflows information was delivered in 2006 by the European Commission (EU), market members were astounded by the low discharges levels versus recompense supply, which ignited an emotional fall in costs. This issue was additionally connected with the powerlessness to bank credits, which we talk about beneath in area 0. Huge cost swings straightforwardly connected with information issues seem, by all accounts, to be for the most part missing beginning around 2006–2007. In any case, states have gone under a few analysis for not delivering convenient and point by point information on individual recompense exchanges and property (de Perthius, 2011). A subsequent issue influencing the EU ETS has been the kinds of balances permitted under the CDM, which we examine in segment 0. Thirdly, the EU ETS has confronted three high-profile instances of market control, two of which were not well defined for outflows exchanging markets. One of these two cases included brokers controlling worth-added charge (VAT) regulations in various nations to cheat legislatures of over €1 billion from 2008 to 2009, while the second included cyberattacks that probably took over €50 million worth of recompenses on spot trades in 2011 (De Perthius, 2011; Frunza et al., 2011). The one significant discussion one of a kind to discharge markets happened when CDM credits recently gathered by the Hungarian government for consistency reemerged the market. Apparently the Hungarian government essentially traded the CDM credits for one more sort of carbon resource under the Kyoto Protocol that is expected to sell. While the trade was totally legitimate under the Kyoto Protocol, it was amazing for some members in the European Emissions Trading System and made the presence of credit "reusing" that would have nullified pertinent carbon decreases and reduced the respectability of the exchanging framework (De Perthius, 2011). After each occurrence of market control, the EC reexamined its principles to resolve the particular issue. For the VAT extortion and CER reusing issues, the Commission has progressively concentrated oversight of the EU ETS, making it harder for people to exploit contrasts between public level regulations and guidelines. The cyberattacks, which constrained a multi-day suspension of spot exchanging, incited the Commission to reevaluate the safety efforts set up by the different trades that work with exchanging (European Commission, 2010). Concerning RGGI, the program's free-market screen has found no significant abnormalities since exchanging started in 2008 (Potomac Economics 2009, 2010, 2011). Market and closeout information is delivered by RGGI consistently, and stipend possessions are recognizable online through the program's CO_2 Allowance Tracking System (COATS) last issue for market conduct has been vulnerability originating from the focal job of strategy. Since outflows markets (barring intentional business sectors) are made by state run administrations, changes in arrangement can possibly decisively influence those business sectors. As the EU ETS enters its third stage, the EC is allegedly considering a deferral in unloading a large number of EUAs, which would almost certainly drive up momentary costs (Allan, 2012; Szabo, 2012). Be that as it may, except if those recompenses are resigned for all time, they in the end would come into the market, stifling future costs. Furthermore, program changes planned to diminish the stock of EUAs might diminish the degree of conviction market members can anticipate from the EC, possibly deflecting financial backers from taking part in the fates market or from banking remittances. Carbon markets in the United States and Australia additionally face significant vulnerability over strategy. RGGI saw a decline in the size of its market when New Jersey pulled out from the program in 2011. As referenced before, Australia's resistance Liberal Party has promised to dispense with that nation's proposed outflows exchanging program assuming it oversees the public authority in 2013 (Australia Liberal Party, 2012). In New Zealand, rules were reconsidered to permit one remittance to be given up for two metric lots of outflows during a progress stage (Fallow, 2009).

3. Lesson: Banking Matters

As indicated over, the principal arrival of genuine EU ETS emanations information in 2006 showed discharges levels diagramming a course well underneath their Phase I cap. Since EUAs in Phase I was unable to be banked and must be utilized somewhere in the range of 2005 and 2007, an oversupply of licenses implied that costs were probably going to drop—and drop they did. In the principal quarter of 2006, spot EUAs exchanged at €25/metric ton. By the last quarter of 2007, spot costs were basically zero, at €0.06/metric ton. This steep drop didn't address the fundamental cost of carbon, nor was it exclusively an issue of restricted information; rather, it mirrored the failure of market members to bank recompenses for use in ongoing stages. Had producers had the option to utilize their excess Phase I credits in Phase II of the EU ETS, costs would in any case have dropped because of insight about a current period oversupply yet absolutely could not have possibly moved toward the close to zero levels seen in late 2007. The reasoning for not permitting banking was the craving to isolate Phase II, which corresponded with the main Kyoto consistence period, from the EU ETS' Phase I time for testing—however the results of this choice were clear. Spot costs for Phase I recompenses moved toward nothing, even while contract prospects costs for Phase II EUAs floated above €20/metric ton (Point Carbon 2012a). The EC, as well as architects of future projects, took cautious note. In current and future periods of the EU ETS, as well as any remaining significant discharges exchanging programs, limitless banking of grants between stages is allowed. An arising question is actually how much banking an emanations exchanging framework can (and ought to) support. Examinations of the bombed Waxman-Markey bill recommended that by 2022 (10 years subsequent to exchanging's theoretical beginning date), firms would have gathered banked credits of 7–12 billion metric lots of stipends, 140 to 240% of the program's 2022 cap of about 5 billion metric tons (Energy Information Administration, 2009). Late EU ETS gauges propose consistence elements, banks, and other market members are banking almost 2.5 billion stipends, about 119% of Phase II's yearly cap for vestige into Phase III (Neuhoff et al., 2012). According to a regularizing viewpoint, we would believe that market members should be also educated as conceivable about the heading regarding future approach and to likewise bank. Be that as it may, assuming policymakers act in the surprising ways noted toward the finish of segment of 0, this might have outcomes. For instance, New Zealand's late choice to permit one recompense to count for two metric lots of outflows successfully parts consistence costs. The opposite would be valid in the event that the EU decides to keep down unloaded stipends. Regardless of whether transitory, these encounters of having the public authority venture into the market will impact the manner in which market members view the inflexibility of the emanations cap and ensuing recompense costs. A significant, unanswered inquiry is how much legislatures should be focused or adaptable despite different occasions.

4. Lesson: Allowance Allocation Can Involve Large Revenues and Distributional Impacts

Each significant emanations exchanging program dispenses at any rate a few free recompenses to weighty industry and, sometimes, power generators. In Phase I of the EU ETS, most NAPs constrained power generators to diminish their discharges more than different areas. Worried about a subsequent ascent in buyer power costs and energized by industry, controllers, and chose authorities distributed the power area an enormous portion of remittances at no charge (Markussen and Svendsen, 2005). In aggressive business sectors, for example, Germany, power generators passed along the open door expenses of these free stipends to their clients, permitting generators to separate leases generally practically identical to their extent of uninhibitedly assigned remittances (Sijm et al., 2008; Ellerman et al., 2010). Set forth plainly, power organizations actually charged clients for grants they got for nothing. This anticipated market result showed that free designation didn't be guaranteed to safeguard customers however had distributional outcomes, and was not heartily gotten by certain eyewitnesses (Gow, 2006; Harrison, 2009). The EC has answered by modifying the program's rules, restricting free portions and expanding the extent of remittances sold at sell off. Since weighty industry faces huge global rivalry, which will in general keep firms from passing on stipend or decrease costs, weighty industry firms will keep on getting significant free assignments. Conversely, power generators will get restricted free stipends, and just in determined countries.

Regardless of notable public money contentions against reserving, state run administrations may likewise look to put rents removed from emanations exchanging projects for specific purposes. In many projects, (for example, in Alberta, Australia, California, and RGGI), state run administrations should put a portion of incomes raised from unloading discharges toward low-carbon energy projects. These ventures incorporate environmentally friendly power projects, carbon catch and capacity (CCS) innovation, and proficiency programs. RGGI states, for instance, contributed north of 60% of their closeout incomes (generally $480 million) toward energy effectiveness and sustainable power programs somewhere in the range of 2008 and 2010. Different incomes from RGGI went toward direct energy help to families, and a little offer went to non-related government projects (Regional Greenhouse Gas Initiative Inc. 2011).

5. **Lesson: Significant Competitiveness Impacts and Emissions Leakage Are Not Inevitable**

Without even a trace of a worldwide system, diminished monetary seriousness of covered areas is in many cases referenced as a critical worry with carbon duties or outflows exchanging programs. Two unmistakable issues emerge in this specific situation. Initial, an outflows exchanging program involves monetary expenses and can rearrange financial action—the two positions and capital—away from controlled areas, especially discharge serious assembling. These expenses and reallocation are monetary issues. Also, a portion of this assembling action might move to unregulated wards, with homegrown outflows decreases returning somewhere else on the planet. As well as intensifying the monetary issues, such discharges spillage is a natural issue. On the off chance that critical, spillage brings into question the fundamental natural reasoning for emanations exchanging programs. Nonetheless, a survey of the (restricted) that's what exact writing demonstrates, essentially for the beginning stages of the EU ETS and RGGI, cutthroat misfortunes and spillage seem to have been little in the couple of areas where it has happened. Ellerman et al (2010) found "no noticed effect" on intensity in the oil refining, concrete, aluminum, or steel areas during Phase I of the EU ETS. Demailly and Quirion (2008) found that Phase I of the EU ETS made just a little loss of intensity in the iron and steel areas. Lacombe (2008) found a comparative restricted influence on the EU refining area during Phase I. An examination of the EU aluminum area by Reinaud (2008) found no factual proof of negative seriousness influences from the program yet noticed that data holes remain. Nonetheless, one study of firm directors proposes bigger cutthroat effects and related spillage in specific areas. For instance, 55% of review respondents in metals producers, and 44% of mash and paper and concrete, lime, and glass makers, expressed they have either moved or are thinking about moving out of the EU ETS consistence zone, while just 14% of different firms expressed they have moved or are thinking about such a move (Point Carbon, 2011). There are motivations to respect such overview results with alert. To start with, the study comprises of a tiny example size (215 respondents for the whole EU economy). Second, respondents may not disaggregate the expense of CO_2 from the general expense of power, possibly prompting an overemphasis on the heaviness of carbon costs. On the off chance that the lower gauges from the EU are exact, these insignificant intensity effects and spillage rates might mirror the unobtrusive GHG decrease targets carried out in the main period of the EU ETS. Also, vulnerability over outflows decrease arrangements all over the planet might postpone firms from pursuing choices on moving to locales where GHG strategies could go into impact after the organizations' migration. These noticed seriousness effects and spillage by and large fall underneath the levels anticipated by certain investigations of discharges evaluating in the United States. Aldy and Pizer (2008) assessed that a \$15/metric ton carbon cost in 2012 would bring about diminished weighty modern creation by 1.6–3.4%. Ho, Morgenstern, and Shi (2008) assessed that a \$10/metric ton cost would diminish the creation of most exchange uncovered weighty industry by somewhere in the range of 1 and 3%, however a few businesses would see diminishes up to 7.7%. A U.S. interagency examination of the proposed Waxman-Markey charge (Interagency Competitiveness Analysis Team, 2009) assessed that the regulation would bring about an expansion in negligible creation expenses of somewhere in the range of 0 and 4%. At last, Fischer and Fox (2009) assessed that a high (\$50/ metric ton) carbon cost could bring about spillage paces of up to 27% for some energy-serious areas. Spillage and intensity issues in RGGI have additionally been a worry. Since RGGI states work in a power market coordinated with non-RGGI states, the potential for spillage in the power area plainly exists. Some exploration has recommended spillage rates going from 28% with \$3/metric ton costs to 90% with \$7/metric ton costs (Chen, 2009; Wing and Kolodziej, 2009; Kindle et al., 2011). Nonetheless, low carbon costs coming about because of a powerless economy and generally low petroleum gas costs might have forestalled broad spillage in RGGI. One 2008–2010 exact concentrate in New York State proposes that recompense costs were excessively low to cause spillage (Kindle et al., 2011).

6. Lesson: Offsets Can Work, But They Are Complex

The Kyoto-laid out CDM and JI components have upheld a huge number of outflows decrease projects in the creating scene, permitting the EU ETS and other exchanging projects to diminish costs while supporting alleviation endeavors in non-industrial nations. We will zero in here on the CDM, as CDM projects have dwarfed JI projects by north of 12 to 1 from 2007 to 2011 (Fenhann, 2012). Precisely which kinds of activities ought to get counterbalanced credits has gotten the most consideration. For balances to genuinely diminish outflows, credits must be given to projects that could not have possibly happened without the counterbalanced credit program. That is, projects should furnish extra emanations decreases contrasted and an existence where the CDM didn't exist. Albeit much exploration shows that the CDM has brought about genuine outflows decreases, an assortment of analysts have portrayed situations where non-extra undertakings got credits under the program (Wara, 2008; Elsworth and Worthington, 2010; Lambert, 2011; Zhang and Wang, 2011). The most hazardous task type in the past has been HFC-23, a refrigerant Utilized in modern cycles that has multiple times the worldwide temperature alteration capability of carbon dioxide (United Nations Framework Convention on Climate Change 2012a). Due to its high an Earth-wide temperature boost potential, projects that decrease HFC-23 get a lot of credits. In spite of a tiny extent of activities, HFC-23 decrease projects got an enormous portion of decrease credits from 2005 to 2006. Lambert (2011) finds that lacking standard estimations made an unreasonable motivating force, empowering HFC-23 producers to briefly build their discharges, permitting them to later lessen HFC-23 result and guarantee a huge number of important credits. Following such examination, the EC casted a ballot in 2011 to prohibit credits from projects diminishing HFC-23 (European Commission, 2011). Also, the CDM Executive Board reconsidered its rules with respect to HFC-23 and other comparable gases, prompting an expanded number of dismissed project proposals. HFC-23 undertakings have gotten far less counterbalanced credits starting around 2006. The quantity of proposed and executed CDM projects has developed considerably, driven by sustainable power like breeze, sun oriented, or biomass. All things considered, in general, issuance of CDM credits has diminished consistently starting around 2007. This pattern probably reflects stricter standards laid out by the CDM Executive Board, as well as the way that many arranged sustainable undertakings have not yet started getting counterbalanced credits. In Phase II of the EU ETS, the use of counterbalances shifted impressively among EU countries, going from a low of 0% in Malta and Liechtenstein to 13 and 18% in Spain and Lithuania individually. Generally speaking, the utilization of CERs for consistency across all EU countries from 2008 through mid-2012 was around 6%, adding up to 456 million CERs. Whenever ERUs procured from JI projects are incorporated, the figures reach north of 7%, or around 555 million worldwide offsets. However, Phase II NAPs permitted EU countries to involve balances for a normal of 11% of their emanations decreases, implying that producers have decided not to utilize their greatest number of counterbalances. Germany and the United Kingdom, the area's two biggest GHG producers, have utilized less than half of their permitted balances for consistency through Phase II. This restricted utilization of counterbalances in Phase II is logically connected with the capacity of firms to bank their capacity to utilize counterbalances into Phase III of the EU ETS. Assuming firms anticipate that remittance costs should rise (or credit costs to fall), they might forgo involving counterbalances in the ongoing time frame. The CDM has likewise introduced an assortment of distributional inquiries because of the related exchange of assets from Annex-I countries to agricultural countries. Through 2011, by a wide margin the biggest portion of tasks and attributes were going to China and India. Truth be told, somewhere in the range of 2006 and 2011, over a portion of every year's CDM credits went to projects in China (finishing out at 75% in 2007) (Fenhann, 2012). Since China is the world's biggest carbon producer, it isn't is to be expected that a huge portion of GHG decrease tasks would stream there. Nonetheless, countries or political partners that accept China ought to focus on an additional severe emanations decrease plan, or consider China to be a contender, may have a problem with the exchanges empowered by the CDM. Most specialists concur that the CDM has effectively delivered genuine emanations decreases, yet whether the tasks are meeting more extensive advancement goals, for example, monetary development or innovation move, stays questionable (Lecocq and Ambrosi, 2007; Olsen, 2007; Sutter and Parreno, 2007; Dechezlepretre et al., 2008; Schroeder, 2009; Popp, 2011). Considering the worries depicted above, and combined with quickly falling costs rules for CDM projects keep on developing. Right now, an assortment of CDM project types face survey from the CDM Executive Board and the EU ETS, where north of 7,900 CDM or JI projects have been supported, and almost 3 billion offset credits have been apportioned (CDM and JI Monitor, 2012; Fenhann, 2012). In the meantime, individuals from the European parliament have called for limitations on CDM credits for specific sorts of activities (McGarrity, 2012).

It is difficult to close a conversation of counterbalances without basically noticing the breakdown of the CDM market toward the finish of 2012. In the wake of staying around €10–15 territory for the majority of 2009, 2010, and the main portion of 2011, CDM costs fell consistently to under €1 in November and December of 2012. This has been attributed to expanded constraints on the utilization of CDM credits in the EU, vulnerability about future interest, and progressively vigorous stock–issues that should be figured out for financial backers to keep on really believing in offset markets.

9.7. THE FUTURE OF CARBON MARKETS: NEW ISSUES

We are currently at a totally different spot than we were a long time back. In the last part of the 1990s, generally learned and partner banter zeroed in on a solitary worldwide exchanging program being planned as the vehicle to address worldwide environment change [21]. The central points of contention were the plan of that program. Today, that type of hierarchical worldwide program appears to be far away, on the off chance that certainly feasible. All things considered, we see a variety of public and, surprisingly, sub-public exchanging programs arising. Additionally, we currently have genuine involvement in these carbon markets. Simultaneously, the eventual fate of public environmental change strategy in the United States—the biggest created country producer and unique hero of outflows exchanging—is dubious. In the event that an exhaustive U.S. strategy arises, it isn't evident whether that approach will be as an emanations exchanging program, an outflows charge, a tradable exhibition standard, or customary guideline. This raises various new issues that definitely stand out enough to be noticed in the past writing and that sometimes we're not completely expected or perceived during the plan phases of existing carbon market frameworks.

9.8. LINKING CARBON MARKETS

Up front in the new conversation of carbon markets is the manner by which, whether, and when various business sectors can be connected so that directed substances in a single ward can involve recompenses or credits from one more locale for consistency, and potentially the other way around (Jaffe et al., 2009). There are an assortment of inspirations for these sorts of linkages: accomplishing worldwide expense reserve funds and gains to exchange, decreasing homegrown market instability, bringing down homegrown consistence costs, making energy for worldwide activity, tending to expected worries of significant exchanging accomplices, and making the system to turn into a net exporter of emanations credits, among others. Simultaneously, there are difficulties to connecting, most strikingly the risk to natural respectability, the need to orchestrate includes (and comparing loss of power over program plan), and the distributional results of higher (or lower) costs. A large part of the connecting work has zeroed in on the common increases to exchange. To be sure, the early examination of the Kyoto Protocol zeroed in on how much less expensive a worldwide exchanging framework would be contrasted with an arrangement of autarkic frameworks (Weyant and Hill, 1999). As homegrown emanations exchanging proposition started integrating highlights like cost covers, extra work started to demonstrate the way that specific elements in a single framework could prompt expanded discharges assuming two frameworks were connected. For instance, Fischer (2003) shows that connecting a framework that is ordered to yield with a common covered framework quite often increments outflows. As connecting conversations have become more genuine, individuals have started to contemplate precisely which highlights must be adjusted to stay away from such issues and which don't (Mace et al., 2008). Practically speaking, linkages might be one-way or two-way (Mehling and Haites, 2011). In a one-manner linkage, credits in a single framework can be utilized for consistency in another, yet not the other way around. In a two-manner linkage, the two frameworks commonly permit different credits to be utilized for compliance. It is helpful to additionally recognize inside every bearing of linkage between the trade linkage choices and how they connect with previously mentioned concerns (perceiving that any purchase linkage choice by one entertainer addresses a sell-linkage choice, unequivocal or understood, by the vender). A buy-linkage addresses the choice by one exchanging framework to acknowledge for consistence remittances or credits made and made available for purchase by another framework. Sell linkages address a certain or unequivocal choice by one purview to permit or urge different locales to involve its remittances or credits for consistency. Worries about natural uprightness commonly emerge in the purchasing framework (Mace et al., 2008). The purchasing framework is the one concluding that the merchant's credits or stipends are legitimate for consistency in the purchasing framework (the selling framework has proactively chosen to involve their own credits for consistency). Worries about blending elements can emerge in one or the other framework, contingent upon who has more power in the connecting exchange; this is habitually a component of the overall market size. Presently, for instance, the EU set the terms for Norway, Iceland, and Liechtenstein to enter the EU ETS, and California's arrangements have affected the plan of Quebec's market, as would be considered normal to interface with California. In the meantime, the CDM as a merchant of credits stays autonomous, albeit the EU has set specific circumstances on the sorts of CDM credits it will acknowledge. At long last, distributional worries will more often than not emerge in the selling framework. For the purchasing framework, connecting brings down costs with a similar natural result—something many projects want. Lower costs commonly mean less reallocation among different market specialists. For the selling framework, connecting raises costs. There are still gains to exchange for the selling framework all in all, however greater costs commonly mean more rearrangement among purchasers and venders inside the selling framework. Therefore, Australia at first wanted to confine worldwide deals of its recompenses—in spite of the net additions from exchange—however it currently plans to sell remittances into the EU market in 2018 (Jotzo and Betz, 2011). One of the really fascinating (and unanswered) questions is the means by which delinking could work. Ongoing occasions in Europe, with Greece considering an exit from the Eurozone, ought to positively make partners mull over what could occur not too far off. Insofar as connected exchanging frameworks keep up with unmistakable units of record, notwithstanding, the better similarity is a fixed cash framework as opposed to a money association. All things considered, apparently to be not excessively tricky ought to delinking be vital.

9.9. NEW INFORMATION AND REVISION

The conversation of connecting and perhaps delinking, as well as the carbon market experience examined in the past segment, features that strategies obviously will be amended and, surprisingly, redesigned over the long haul. The EU's accounted for thought of keeping down stipends in Phase III and the New Zealand choice to permit one remittance for each two metric lots of emanations during 2010–2012 are great representations. Not at all like traditional guideline, where the monetary outcomes of strategy modifications are restricted to influences on the worth of actual venture, corrections to carbon market rules (or carbon charges, so far as that is concerned) influence the worth of monetary liabilities. They additionally influence the worth of monetary resources–government receipts or, on account of free portion, the worth of future free designations. Banking arrangements in most carbon exchanging programs lead these impacts to happen a whole lot earlier: While strategy changes may not produce results for a very long time, any future change to recompense costs will promptly influence current costs. A normal cost decrease later on implies current stipends are less important to hold while a normal cost increment implies the reverse. While alluring according to a unique productivity perspective, such changes in market esteem suggest monetary effects and outcomes whether or not the thought about change occurs. Assuming that the holders of the recompenses will more often than not be the very same individuals who face consistence commitments, the net impact of changes in market worth could be moderately little, as the market worth of remittances will vary alongside the expense of their future consistency commitment. In any case, precisely the way that remittances are esteemed on accounting reports can make issues in any event, for these businesses obviously, assuming people without commitments are holding stipends, the outcomes could be very emotional. It is hazy whether such effects on theoretical possessions ought to be considered to be fair or out of line, however there are clear effects. What do these market influences suggest for the more extensive issue of strategy correction? To start with, strategy update actually should occur. One of the main traits of environmental change is vulnerability about influences. Combined with the association of strategies in various wards, corrections to carbon market approaches are crucial for long haul proficiency (Murray et al., 2009). While business sectors and impacted partners might need conviction, states can't ensure sureness where it doesn't generally exist. Second, legislatures ought to make progress toward straightforward and methodical approach corrections. Numerous administration organizations, from national banks, to controllers, to courts, go with choices consistently that have huge market outcomes. Administrative offices, courts, and governing bodies all need to seek after market-touchy choices in a manner that permits all market members equivalent admittance to data as well as early notification of the arrangement and timing of the choice cycle. Here the EU ETS has as of late been scrutinized (Allan, 2012; Szabo, 2012).

At last, and generally applicable to the subject of strategy plan, there might be ways that emanations exchanging approaches can be made more manageable to important modifications. For instance, one administrative proposition in the United States would have carried out a quite certain timetable for occasional five year surveys and updates, with official accommodation of proposals not long after the consistence year closes and facilitated Congressional activity inside six months. Another choice is to place these choices under the control of an oversight substance, like a focal bank. Such an element would be answerable for intermittent audits and changes to as far as possible or different guidelines and would have the adaptability to do so deliberatively and outside the unequivocally political circle of assemblies. Given encounters with national banks and money related arrangement, this approach has some allure. In any case, it would be more difficult to adopt this strategy with environmental change in light of the proceeding with difference of perspectives about the suitable degree of reaction, even among specialists.

9.10. ALTERNATIVE POLICIES AND COMPARABILITY

The ending history on discharges exchanging the United States features the way that we are going into a world not just of decentralized, base up outflows exchanging systems with fluctuating standards yet one which a few wards might save emanations exchanging by and large and seek after discharges assessments or more customary guideline. In late 2012, for instance, there was a surprising juncture of interest in carbon expenses to address environmental change and the thriving U.S. shortage (Chemnick, 2012). South Africa has additionally shown interest in a carbon charge (RSA 2010). In the interim, strategy related emanations decreases in the United States throughout recent years (and possible in the close to term) have emerged from more tight guidelines on vehicle mileage and tailpipe GHG discharges, sustainable power limit increments related with government and state appropriations and orders, and new power plant outflows guidelines under the Environmental Protection Agency. This variety of strategy approaches was not completely startling. Under the Kyoto Protocol, there is no necessity to carry out a homegrown discharges exchanging program. The EU ETS, which sits at the zenith of the carbon market conversation, just covers generally 50% of European emanations, with customary guideline or duties utilized somewhere else (various European countries additionally demand an assessment on GHG outflows). Furthermore, in the United States, where worries about seriousness overwhelmed the discharges exchanging banter, the Waxman-Markey bill didn't try to require exchanging accomplices to have an emanations exchanging program, just a "broadly enforceable and economy-wide ozone harming substance outflows decrease responsibility for that country that is actually that tough of the United States". What this promptly raises, notwithstanding, is the need to quantify the likeness of approaches. Equivalence is significant for carbon markets in light of the fact that, among exchanging programs, similarity is important for locales to consider connecting. Also, among the more extensive set-up of arrangements, similarity is important to try not to heighten worries over intensity and outflows spillage that could compromise the manageability of emanations exchanging. The Kyoto Protocol addressed this issue by having nations arrange concurred focuses for each other in view of a 1990 pattern. During the lead-up to the Copenhagen Accord, there was extensive discussion over this as to different nations' pledges. Most conversations check out at discharges decrease endeavors in one of five ways: (1) outflows decreases versus a memorable pattern (percent decreases contrasted with 1990, 2005, and so forth); (2) emanations decreases versus a the same old thing gauge (e.g., percent decreases contrasted with figure levels in 2020); (3) decreases in outflows force (percent decreases in emanations per unit of GDP, energy use, or power age against an authentic standard or future projection); (4) decreases in discharges per capita; or (5) the acknowledged carbon cost. Nonetheless, there is no settlement on which metric is ideal. Many raise down to earth issues (e.g., transformation among monetary standards or computation of the same old thing estimates), and various measurements yield emphatically various perspectives. This inquiry of likeness is possibly intensified while assessing real execution of strategies and their results rather than economy-wide discharges vows. That discussion is just barely starting as nations set out on dealings over another environmental change understanding, one in light of the 2011 Durban Platform for Enhanced Action, and as they wrestle with homegrown partners who are often worried about whether other significant producers, especially their monetary rivals, are embraced their reasonable portion. Dissimilar to the circumstance in

Kyoto, where the component of likeness was a somewhat tight scope of deviations from 1990 outflows levels, future similarity conversations will without a doubt be significantly more perplexing.

9.11. INTERNATIONAL NEGOTIATIONS

What does this suggest for future worldwide dealings concerning carbon markets? Not at all like prior, Kyoto-style exchanges that zeroed in on a grouping of hierarchical, bigger to smaller discharges exchanging issues—public outflows covers, exchanging rules, and further subtleties like the CDM—the new, post-Durban discussions will fundamentally zero in on the devices for a bottom-up approach. From one perspective, another arrangement should uphold worries over similarity and straightforwardness of exertion. Those nations chasing after and as of now taken part in carbon markets will need confirmations that different purviews will do their reasonable portion. This is significant in the event that help in those nations for carbon markets is to proceed, and such confirmations will direct future reactions to intensity concerns. Then again, another arrangement should zero in on ways of offering institutional help for business sectors themselves. Beside prodding the formation of the EU ETS and other exchanging programs, the CDM is seemingly perhaps the main commitment emerging from the Kyoto Protocol. Regardless of the different provokes it keeps on confronting, the CDM has an institutional construction that permits it to develop and benefits essentially from the believability stood to it by its job in the concurred global engineering. The exchanges could search for ways that a more extensive assortment of country commitments can be upheld. For instance, a few emerging nations might need something like "model guidelines" for laying out a homegrown exchanging program that would possibly connect to created country programs previously using the CDM. While rules for carbon markets and other decrease projects can and may arise naturally without an anchor in peaceful accords, making model standards could be significant, especially for the numerous nations that will be excessively little to seek after an altogether altered approach. There are likewise inquiries concerning the fate of the CDM itself. Choices in December 2012 will restrict future admittance to the CDM to nations taking an interest in the following stage (2013–2020) of the Kyoto Protocol. This approach directs the CDM away from a job in a decentralized worldwide carbon market by restricting its significance to the subset of Kyoto members. To accomplish proficiency, future exchanges ought to set out open doors for linkages, not obstructing them.

9.12. SUMMARY

The overall messages from this review are that carbon markets are sizable, they have been extending instead of contracting, market rules are advancing and not static, and another engineering in view of a more natural, base up worldwide plan brings up a set-up of new issues. A worldwide emanations exchanging framework has not arisen as the key part strategy answer for environmental change in the manner it might have been imagined in the mid-to-late-1990s. Other arrangement instruments—carbon charges and customary guidelines—are being executed or considered in an assortment of areas and wards. The plan of carbon markets is profiting as a matter of fact. Experience with bonus benefits from free recompense distribution has prompted an expanded utilization of closeouts. Purviews are figuring out how to deal with market-touchy data in a more straightforward and deliberate way, yet there is still headway to be made. Endeavors to direct high and low costs are giving illustrations on what works and what doesn't while additionally making the basic point that costs matter. Maybe in particular, we are seeing that carbon recompense exchanging can uphold emanations decreases and convey market messages for future speculation. Notwithstanding, the strength of those signs for future speculation depends on trust in the discharges market, the hidden administrative structure and its severity, and the more extensive venture environment. The advancing idea of carbon advertises and related plan changes suggest that trust in the market can't be 100%. Legislatures can't give conviction where it doesn't essentially exist. Looking forward, notwithstanding, specialists should be more clear and all the more efficient about arrangement amendments and perceive the ensuing effects on market cost, market members, and future market certainty. Among the many issues confronting markets from now on, the development of various discharges exchanging programs has put upfront the subject of how, whether, and when these projects will be connected together. While an assortment of inspirations drive interest in connecting, and there are an assortment of ways of making joins, three key worries have restricted connecting up to this point. Purchasers will quite often be worried about natural honesty, as the purchasing framework is laying out that bought stipends are substantial for consistency in their framework. The essential harmonization of specific plan includes likewise implies that either framework is surrendering some sovereign control. Maybe in particular for merchants, selling frameworks will by and large see expansions in recompense costs, with possibly unfriendly distributional outcomes (in spite of in general gains from exchange). Hence, linkages among exchanging frameworks have continued somewhat leisurely. What is the job of global talks with respect to carbon markets in this arising, base up world? One job is to resolve the issue of likeness among various exchanging frameworks as well as among emanations exchanging, charges, and customary guidelines. Similarity among exchanging frameworks upholds connecting, and likeness all the more by and large can help try not to raise serious concerns. The last option concerns, connected with the two emanations and financial spillage, address perhaps the best test to the drawn out supportability of carbon markets. Notwithstanding similarity, it will be valuable to investigate how worldwide organizations can all the more straightforwardly support carbon markets in a more decentralized system. The CDM is a huge commitment; maybe there are others. Fifteen years after the marking of the Kyoto Protocol and the making of the principal vehicle for carbon showcases, the Kyoto model of a hierarchical, worldwide carbon exchanging framework is basically finished. Carbon markets are not, in any case, finished. The test presently is to sort out how they can function in a significantly more complicated—however obviously more practical—world of comparability among different trading systems as well as among emissions trading, taxes, and traditional regulation. Comparability among trading systems supports linking, and comparability more generally can help avoid escalating competitiveness concerns. The latter concerns, related to both emissions and economic leakage, represent one of the greatest challenges to the long-term sustainability of carbon markets. In addition to comparability, it will be useful to explore how international institutions can more directly support carbon markets in a more decentralized regime. The CDM is a significant contribution; perhaps there are others. Fifteen years after the signing of the Kyoto Protocol and the creation of the first vehicle for carbon markets, the Kyoto model of a top-down, global carbon trading system is essentially over. Carbon markets are not, however, over. The challenge now is to figure out how they can work in a much more complex—but clearly more realistic—world.

CHAPTER 10

STRATEGIC REPORTING AND COMMUNICATIONS

CONTENTS

10.1. REPORTING AND COMMUNICATIONS

Sustainability reporting is a fundamental initial phase in the process by which businesses make a move to make a feasible worldwide economy; this conviction has been supported by GRI's main goal. Be that as it may, reporting is not static-it advances, through improvements like rising worldwide difficulties, innovations, and changing partner interests- and organizations need to think about this.

The Corporate Leadership Group (CLG) on Reporting 2025 was met by GRI to advance the worldwide dialog on the fate of sustainability transparency and divulgence, and to analyze how the act of reporting is developing.

Around 13 forward-looking organizations met up to research the patterns they predict will be more important later on. From the 17 patterns GRI displayed at the Sustainability and Reporting Trends Forum held in November 2015, the four were researched and better comprehend the impact they would have all alone organizations and sustainability reports (Figure 10.1).

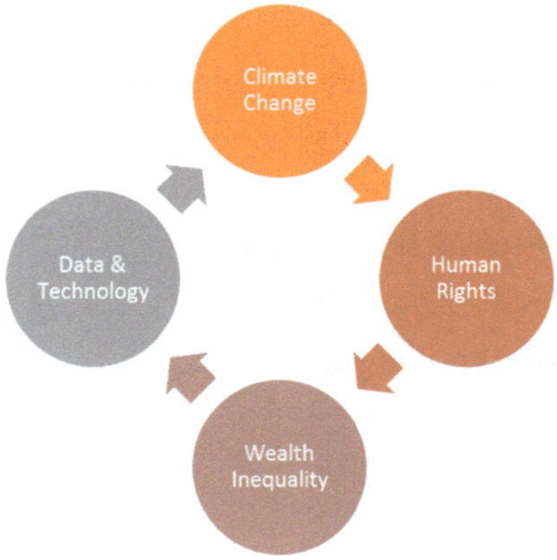

Figure 10.1: Reporting and communication areas.

Source: GRI (2017).

10.2. EXPLORING FUTURE TRENDS

To increase new bits of knowledge and a more profound comprehension of the potential impacts of these patterns all alone organizations and sustainability reports, the four patterns were dissected in four angles.

Diverse specialists were welcome to impart to the gathering their considerations on the fate of reporting, and sustainability gave content and encouraged the talks. In the Labs, the gathering adopted an organized strategy to investigating the patterns, covering the accompanying four viewpoints: (Figure 7.2).

Figure 10.2: Reporting process.

Source: GRI (2017).

10.3. EXAMPLE OF CLIMATE CHANGE

10.3.1. Current Landscape

Environmental change is currently the worldwide plan and is a standout amongst the most develop yet advancing patterns in sustainability and corporate reporting. In 2016, the World Economic Forum named the inability to alleviate and adjust to environmental change the top worldwide hazard regarding sway. The environment is highlighted noticeably in the 17 sustainable development goals (SDGs), which are now accumulating noteworthy intrigue and early take-up. Many transparency businesses as of now unveil data on environmental change; the Organization for Economic Co-operation and Development (OECD) and the Climate Disclosure Standards Board (CDSB) took supply of environmental change divulgence in the G20 nations to locate that corporate reporting practices are enhancing quickly, with a practice that was considered spearheading in 2010 getting to be noticeably standard in 2015.

The point of interest made at COP21 in Paris in December 2015 brought about:

- A more tightly atmosphere strategy condition;
- A long-haul objective being set for worldwide de-carbonization;
- Momentum in financial specialist enthusiasm for the issue, so far playing out by segment lines;
- The sureness that administrations cannot accomplish worldwide targets alone.

The following stride of taking a load of governments' Intended Nationally Determined Contributions (INDCs) to environmental change is in progress.

Various related activities have additionally risen in 2016:

- The OECD has been appointed to evaluate the past responsibility of a $100 billion store by 2020 to help developing nations adapt to environmental change;
- The Breakthrough Energy Coalition of innovation leaders is giving $1 billion to development to people in the general research pipeline;
- 120+ organizations have focused on creating science-based targets.

Corporate authority on environmental change should be driven in various regions. The top variables driving organizations to address environmental change are:

- Using elective and sustainable power source;
- New advancements/development;
- Setting targets and having an atmosphere methodology;
- Communicating and supporting on atmosphere;
- Supportive administration;
- Leadership from the top;
- Committing and conveying genuine outcomes;
- Reviewing products and administrations for atmosphere impacts.

"The Bank's commitment to the Paris Ascension is an essential piece of our need setting. Atmosphere related, green fund, green securities are issues where the Bank has been centering for a considerable length of time, and positively that concentration is not going to leave."—Parminder Plahe, European Investment Bank (EIB).

Besides, environmental change is being regarded money related: The Financial Stability Board presented the Task Force on Climate-related Financial Disclosures to create intentional predictable atmosphere-related monetary hazard exposures for organizations to give the correct data to speculators, moneylenders, backup plans, and different partners.

10.3.2. Implications for Business

There was clear agreement in the gathering that it is not a matter of if business ought to or can follow up on environmental change however how and how quickly they convey change. The accompanying focuses were exhibited in the Lab for the CLG to talk about:

- Companies are arrangement suppliers: they are relied upon to be a piece of the arrangements, from new energy models to efficiencies in the generation and dispersion of products.
- Climate change is interrelated with different issues: how does environmental change communicate with and affect different issues, for example, human rights?
- Clear, hunger, science-based targets are required: new indicators on environmental change impacts and related externalities are developing yet deficient even with concurred worldwide cutoff points.
- Linking to INDCs is vital: more prominent organization and nation engagement is relied upon to be required with the Paris understanding.

"Environmental change impacts horticulture. For our sourcing, the test is how to draw in ranchers and those included in rural creation on environmental change. Our report ought to be more unequivocal, more realistic, and clearer in its objectives to make it more conceivable for our own particular partners, particularly for our accomplices in horticulture."—Sol Beatriz Arango, Grupo Nutresa.

10.3.3. Examples: Business Addressing Climate Change

Cases of how organizations incorporate atmosphere-related data in their sustainability reports:

- CEMEX has detailed widely on environmental change, ordering advancement, and future course into an independent position, which incorporates advance against targets and flags a move to inner carbon valuing.
- Power organization CLP set an objective in 2007-which still stands-to lessen the force of carbon outflows by around 75% by 2050.
- Food and refreshment organization PepsiCo is working towards a flat-out lessening in ozone harming substance emissions over the business and utilizes the focal point of the PepsiCo esteem bind to demonstrate the limits of this material issue.
- EIB, the world's biggest multilateral lender, has focused on straightforward and nitty-gritty giving an account of the atmosphere advantages of its speculations, by catching the GHG emissions of 95% of its venture ventures while screening out activities whose GHG outflows surpass a stringent edge level.
- Renewable energy bunch Gamesa plans to accomplish carbon lack of bias by 2025. This test includes singular duties on the estimation, lessening, and additionally counterbalancing of the CO_2 starting from the Group's immediate and roundabout discharges from 2015 to 2025.

To see other driving cases, see the GlobeScan/Sustainability 2015 Climate Survey, which recognizes driving corporate environmental change objectives. One territory where there is wide accord is setting aggressive targets: RE100 is turned out to be an intense activity, through which a developing rundown of organizations has focused on 100% sustainable power.

10.3.4. Priorities and Next Steps

Priorities for exploring and improving their reporting on climate change:

- Contextualize and disguise the test of environmental change more: Learn what you can from best practice yet guarantee you think about your materiality examination and the parts of environmental change most important to your division and association and enhance information, measurements, and reporting as needs be, for example:

- if the accentuation is on your inventory network, your need may comprehend the dangers in your store network and computing your extension 3 emissions.
- if your products and administrations impact the energy utilization and discharges of others, their plan and conveyance might be generally intense.

- Aim to comprehend the interconnections, or nexus, of environmental change and different issues.
- Set and accomplish long-term (past 2020), science-based targets (endeavor to get at 'what's coming to you'), which likewise separates organizations.
- Determine how to adjust the interests and exercises of transparency businesses on governments, given national anticipates environmental change and the multinational way of many reporting businesses.
- Develop a reasonable and compact story of the helplessness of your association in light of environmental change and the common test it presents.
- Focus on conduct change: what will people and businesses need to do any other way to address the difficulties around energy utilization and emissions? "In a resource-constrained world, CEMEX understands the importance of reducing emissions and the need to quickly incorporate clean energies in the elaboration of our product. We understood we had to come to solutions quickly to contribute to a cleaner product and decrease our footprint."—Martha Herrera, CSR Director, CEMEX.

10.4. EXAMPLES OF HUMAN RIGHTS

10.4.1. Current Landscape

Human rights insurance is a key, and developing, societal concern, especially in light of ongoing clashes (e.g., in Syria) and relocation and displaced person streams. For the greater part of the advanced business has made human rights manhandle still happen. Many organizations are as yet pondering worries in their supply chains, specifically-be the unprotected transient workers in the palm oil or fish supply chains in Asia, the worry for Syrian outcast youngsters working in clothing production lines in Turkey, or present-day bondage in the UK. We see significant improvements with respect to human rights moving towards national execution and responsibility:

- Human rights are implanted over the SDGs: the Danish Institute for Human Rights proposes we can "... utilize human rights as a driver for understanding the SDGs... and utilize the SDGs to acknowledge human rights..."
- Various countries-presently including the UK, Italy, Spain, and Colombia-have created National Action Plans particularly for human rights.
- The UK's Modern Slavery Act was viewed as historic point enactment, influencing an expected 17,000 organizations: huge UK-based firms are required to distribute confirm on their sites showing that their operations are not dependent on universal subjugation.

10.4.2. Drivers of Corporate Performance on Human Rights

Phil Bloomer, Director of the Business and Human Rights Resource Center, specified two patterns that will be basic to business operations and show the interconnectivity of the patterns:

1. **The Ascent of Hyper-Disparity and Chance of Shared Thriving:** Business assumes a crucial part in the route in which riches are made, additionally in how riches are conveyed. Regarding and securing human rights is major to the equivalent conveyance of assets.
2. **The Compounding Biological Emergency:** The need to move to a low carbon economy makes a chance to discover arrangements that added advantage the most powerless-those frequently left unprotected earth or naturally likewise regularly encounter an absence of human rights and social insurance, particularly in long and complex supply chains.

10.4.3. Implications for Business

Desires of corporate providing details regarding the numerous features of human rights are developing and the issue may quickly develop:

- Human rights due persistence are presently the normal least.

- Benchmarking and positioning will empower simple and fast examinations between businesses.
- Investors, rating offices, and controllers are progressively looking for this data.
- Greater required straightforwardness and due steadiness: This is obvious all around with requests from common society and expanding arrangement and control.
- An emphasis on work rights as basic: Modern subjection is another type of extreme work manhandle, and is prompting a more extensive development from attention on review and consistency to due tirelessness and cooperation.
- Stakeholders are stressing the requirement for least living wages, hence influencing brands and unions, modern aggregate understandings and possibly making compensation pre-aggressive. Well-being and security, the flexibility of affiliation, and the privilege to aggregate haggling likewise remain concerns.
- Conflict over common asset riches: Land rights are progressively debated and, as water shortage declines, questions are required to develop. Extraordinary air contamination is likewise picking up consideration. As the world moves to low carbon economies, there is support to guarantee that the move is quick and reasonable. At long last, corporate duty and assessment asylums have a part to play to guarantee expense is paid where asset riches are separated.
- Emerging zones of solid corporate promotion: various companies have talked and acted to end segregation by sex or sexual introduction, for example, while in a few divisions, for example, assembling, extractives, and innovation-support for the regard of fundamental rights all through the inventory network remains a need.

"Ensuring that compensation and remuneration programs do respect and improve minimum wages established in different markets and countries, or reducing the gender wage gap becomes a key challenge for multi-located companies like ours."—Isaac Ruiz, CSR Director, Gamesa.

10.4.4. Examples: Business Addressing Human Rights

Cases of how organizations incorporate human rights-related data in their sustainability reports:

- ICT organization Telefonica concentrates on the human rights issues around computerized trust, which incorporate security and security, the flexibility of expression, and youngster insurance. Telefonica is chipping away at the execution of its 10 Guiding Principles on Privacy and Freedom of Expression.
- Grupo Nutresa has a structure of activity-a reasonable independent correspondence on where the organization remains on human rights and how they conform to their human rights arrangement.
- Renewable energy organization Gamesa and Industrial Global Union consented to a worldwide structure arrangement (GFA) including a pledge to regard fundamental social rights, human rights, and common freedoms, level with circumstances, justify, and non-separation.
- Indian iron and steel maker JSW trusts a various workforce is a key element for a creative biological community. They exhibit the breakdown of the workforce by age, sex, classification, level, and turnover.

10.4.5. Priorities and Next Steps

- Better comprehend human rights, not just as a material issue for some organizations, however a greater comprehension of striking nature and which rights are in danger of the most serious effects through the organization's exercises and business connections.
- Consider doing this through human rights impact evaluations.
- Consider receiving the UN Guiding Principles on Business and Human Rights Reporting Framework.
- Building mindfulness and spreading learning all through an organization and its inventory network is crucial.
- Companies are engaged with investigating how their HR offices can help serve human rights.
- Integrating regard for human rights all through the production network is viewed as the greatest test.
- Companies are additionally intrigued by stages to share store network data, as a huge number of providers in any one inventory network cannot be checked or inspected by a solitary association viably.

- Multi-partner coordinated effort for aggregate activity around a range of human rights, regardless of whether ware particular or a neighborhood area, is testing yet worth handling.
- Supply chain chance investigation work and inspecting is an initial step to distinguishing dangers in human rights.

10.5. EXAMPLES OF WEALTH INEQUALITY

10.5.1. Current Landscape

As per Oxfam, the wealthiest 1% on the planet now have more riches than whatever is left of the world consolidated. In the sustenance and drink industry, for instance, 60% of ranch workers live in neediness. Little scale ranchers create 80% of the world's nourishment, yet together with horticultural laborers, they make up the lion's share of the ravenous individuals.

The OECD characterizes riches imbalance as "monetary disparity… the distinction in how resources, riches, or pay are circulated among people as well as populations." It is likewise portrayed as the gap between rich and poor, pay imbalance, riches divergence, riches, and pay contrasts, and the riches gap. Tending to riches disparity is progressively observed as a crucial essential of a feasible economy, with various key partners and specialists concentrating on the issue and how it interfaces with get-to-monetary assets, instruction, wellbeing, and so forth. With the ascent of hyper imbalance and every one of the dangers it presents, there is a significant chance to concur and re-adjust on what is more equivalent and shared success.

Examining the sustainability reporting of a few organizations, it is clear that riches imbalance is infrequently tended to straightforwardly as a pattern or some portion of the more extensive working setting impacting business. There are signs that more organizations are perceiving the potential and desire for them to impact financial difficulties through particular business exercises, for example, work, wage levels, and proportions, and neighborhood sourcing.

10.5.2. Drivers of Corporate Performance on Wealth Inequality

As indicated by Ioan Nemes, Policy Lead CSR, and Sustainable Agriculture at Oxfam Novib, there are two key measurements of driving worry for riches imbalance and picking up organizations' consideration:

- An expanding offer of the estimation of our worldwide economy is going to capital, as opposed to work, which addresses the difference between lower-wage specialists and high riches people. Many expense frameworks do not guarantee the reasonable dispersion of riches from capital or benefit-producing exercises to whatever remains of society-in creating markets, as well as in developing ones.
- Inequality is reflected in the way that power, benefit, and impact are utilized as a part of the linkages in supply chains (delineated by the Oxfam Behind the Brands battle).
- What is more, there are empowering signals from worldwide organizations driving interest and consideration on riches balance:
- UN Sustainable Development Goal 10 decisively concentrates on decreasing disparity with particular targets including continuously accomplishing and supporting salary development of the last 40% of the population at a rate higher than the national normal by 2030.
- The World Economic Forum additionally recognized riches disparity as a top pattern in 2015 and attempted a local investigation of the issue to better comprehend which imbalance-related angles are needs in which districts (e.g., defilement, instruction, urbanization), which can better educate corporate comprehension of key markets. Implications for business.

Oxfam sees different difficulties ahead for the business identified with riches disparity, including:

- Radically expanding the share of significant worth caught by laborers and little scale makers.
- For example, accomplishing living wages for workers (and living earnings for little scale makers).
- Eliminating financial gender equality imbalance and sex segregation.
- Tackling the race-to-the-base on open administration to pull in speculation.
- Calling for a conclusion to the period of duty shelters.
- Breaking market fixation and tending to the unequal conveyance of energy.

- For instance, new plans of action ought to be quickly settled to share esteem more impartially in supply chains.

Examples: business addressing issues related to wealth inequality.

Cases of how organizations incorporate human rights-related data in their sustainability reports:

- Ferrero shows expanding straightforwardness and broad reporting at the national level (for key nations), exhibiting through transparency how business operations affect nearby economies. The report by the nation on wage proportions (and kind of laborer), nearby administrative contracts, and neighborhood provider spend.
- Mosaic presents giving an account of its worldwide group speculations-which are organized and arranged as identified with nourishment, water, and nearby-making utilization of a microsite, Mosaic's Giving Map.

10.5.3. Priorities and Next Steps

Needs to explore and enhancing their providing details regarding riches imbalance:

- Start some place-take a gander at your own operations and employees with a social rights point of view and after that consider extending the limits of this evaluation to incorporate the store network. Ask yourself: are there aberrations in the path in which the organization values individuals and exercises in the inventory network?
- Seek more data and detail on where the organization has made group-level ventures and following going ahead.
- Consider how your organization as of now does, or can, impact access to fundamental products and administrations (e.g., sustenance, water, energy) on a worldwide premise or in the neighborhood groups in which you work.
- Explore effect and results-based estimation, regardless of the difficulties of middle people in supply chains, diverse purviews, distinctive exercises, and so on.
- Consider what your organization's most material effect identified with riches imbalance might be and create effect-related key performance indicators (KPIs).

10.6. EXAMPLES OF DATA AND TECHNOLOGY

10.6.1. Current Landscape

Information and innovation are a piece of a steadily developing field of improvements. The CLG watches a speeding up of the measure of information being made accessible, incorporating into constant: more information has been made in the previous 2 years than in the WGAP history of humankind. There are more methods of utilization of information crosswise over esteem chains, including systems of systems and a web of things.

However numerous corporate columnists indicate a requirement for more connection amongst information (and innovation) to comprehend impacts. For sure, for the world to meet SDG target 12.6 is to "empower organizations, particularly expansive and transnational organizations, to embrace sustainable practices and to coordinate sustainability data into their reporting cycle"—we should better comprehend and utilize information. The UN knows this and requires an information insurgency to make information bolster improvement and exhibiting why it truly matters. We know certainly that better accumulation of and access to information (frequently through innovation) empowers transparency that depends on increasingly and better information. Better information empowers partners to comprehend an organization's execution and thus settle on better choices.

Innovation can likewise empower partners to get to information more successfully. We watch corporate reporting utilizing information and innovation from multiple points of view: how organizations are straightforwardly exhibiting and imparting information remotely, how innovation can empower information openness and introduction, and how information can educate partner basic leadership.

10.6.2. Drivers of Corporate Execution on Information and Innovation

Pavan Sukhdev, Founder, and CEO of GIST Advisory joined the Lab on information and innovation as a specialist to extend individuals' reasoning on information and particularly what information organizations may as of now have on the off chance that they choose to esteem it, displaying a multi-capital esteem system. As far as he can tell with organizations, sparking alight by gathering information on human, physical, regular, and social capital can accomplish a multi-dimensional feeling of significant worth that is reflected in sustainability contemplations being better coordinated into organizations and reporting. Understanding these perspectives better- and for a few, connecting a money-related incentive to them-will empowers organizations to recognize potential territories to enhance, to expand positive effects, diminish negative effects, and at last increment business esteem.

10.6.3. Suggestions for Business

With regards to corporate detailing, information, and innovation are regularly observed as an open door and a test in equivalent measure.

10.6.4. Challenges

- Securing adequate internal purchase in.
- Promoting the way of life: making mindfulness for good utilization of the internal frameworks that convey high caliber, practically identical information.
- Availability of delicate and classified information.
- Need for more scientific instruments to better comprehend information.
- Manually ordering information crosswise over spreadsheets or decentralized frameworks, or transitioning to new, more imaginative frameworks to oversee information and data.

10.6.5. Opportunities

- Online transparency is an innovative device to impart information adequately.
- Embedding sustainability information (counting gathering and auspicious accommodation) into targets and execution administration frameworks.
- Being proactive and keeping associates educated around the procedure and results.
- Monitor- and give criticism circles to information suppliers- both the positive and the negative effects of sustainability issues followed.
- Understand the flow and different requests on similar information to enhance the data channels and lower the weight for associates, where conceivable.

10.6.6. Cases: Business Tending to Information and Innovation

Cases of how organizations exhibit information in their detailing, concentrating on internet reporting, intelligent transparency that utilizations information to give the client a chance to pick how and what they read, infographics that present mind-boggling or a lot of information in a simple to peruse design, and regular execution refreshes:

- TIM presents its 2015 online report also outlined and takes into account a simple route to its key concentrate on social advancement, natural assurance, and computerized culture and presents the report with a video.
- CLP opens up a few datasets to enable searchers to interface with and redo their perspective of CLP's information, and it covers money-related, operational, natural, and social and administration execution information.
- Grupo Nutresa presents complex information in a simple to peruse design by outwardly mapping the organization's exercises and commitments crosswise over many issues to particular SDGs.

10.6.7. Needs and Next Strides

Needs to explore and enhance their transparency using information and innovation:

- Use information and innovation to approach partners in various ways.
- Use innovation to facilitate the gathering of information by moving towards more advanced and brought-together information frameworks.
- Explore and better comprehend the requirements of various partner gatherings and tailor the introduction of information in a like manner.
- Present data to partners (e.g., infographics, report microsites) more successfully.

10.6.8. Future Focus

- Greater desires: Overall, organizations are encountering more prominent weight from partners on sustainability issues, as confirmed by improvements, for example, expanding direction.

- Explore and test partner enthusiasm for transparency in various arrangements and frequencies and consider how reporting businesses may deal with this.
- Closely screen and research the four patterns (environmental change, human rights, riches disparity, and information and innovation), particularly in the less developed regions of human rights and riches imbalance.
- Convey the estimation of the development towards GRI Standards in connection to new or rising ranges: GRI can give more prominent clearness and direction on divulgences and guidelines, however, ought not to be prescriptive on the transparency procedure.
- Reset targets: Companies are resolved to return to their objectives; how might they be made longer-term and drive sustainable results? Will more effect and results-based estimation be actualized?
- Provide direction to organizations on targets. Would it be advisable for them to be science-based? Setting based? Speak to a what is coming to the organization of a test or effect?
- Country-level and nearby focal points: There is a craving to search out more national level and neighborhood information over these patterns-on wages, on striking human rights, to line up with government anticipates atmosphere, and so on.
- Explore promote direction on the most proficient method to report adequately on expense rehearse crosswise over various wards, which governments can bolster this.
- Support inquiring about, benchmarking, and direction on ranges, for example, least wages and living wages in various locales or human rights benchmarking to help organizations look at and report.
- Deepen the comprehension of consequences and effects: The need to better comprehend dangers and effects is foremost, regardless of whether that is scope three emissions, leading human rights effect appraisals, or investigating how riches imbalance plays out.
- Support business transparency enthusiasm when estimating effects and consequences.
- Improve information: There are enhancements to be made to internal information accumulation procedures to enhance transparency against these patterns and bolster better understanding.

- Pursue further or more profound arrangements (e.g., GRI with different systems) so information is gathered once for various purposes.
- Encourage peer learning and sharing of best practices.
- SDGs matter: There is a solid accord that the understanding of the SDGs is critical and will affect corporate sustainability motivation.

10.7. TRENDS OF GLOBAL REPORTING

10.7.1. History of Reporting

- 1994–2000 Birth of reporting:
  - First movers;
  - Origins of GRI.
- 2000–2006 Era of innovation and rapid growth:
  - GRI G2 and G3;
  - AA1000AS.
- 2006–2012 Era of calm and steady growth:
  - GRI only game in town;
  - Shared sense of best practice.
- 2012–2015 Era of disruption and questioning:
  - G4, SASB, IIRC;
  - Regulation and customer scrutiny.
- 2015-plus Era of digital technologies.

10.7.2. Three Key Global Reporting Trends

1. **Materiality:** Companies are focusing on sustainability management and disclosure through reporting on topics of highest priority (and most relevance) to the company and stakeholders.

2. **Integration:** Several companies are integrating sustainability management into the core of the business, and producing integrated reports as an output.

3. **Interactive:** Companies are thinking about the difference between report and reporting; a report is static; reporting is more dynamic: multi-channel, ongoing, and/or social.

10.7.3. The Different Dimensions of Materiality

Multiple criteria can be used when applying the materiality principle and prioritizing sustainability issues. Here are some of the most common examples:

- Importance to business success;
- Importance to sustainability and/or stakeholders;
- Changing importance over time;
- The ability of the company to influence;
- The extent of current management.
- **Materiality Analysis:** Aims to identify the societal and environmental issues that present risks or opportunities to a company while taking into consideration the issues of most concern to external stakeholders. The process of labeling material topics in a company requires a comprehensive framework that methodologically identifies and prioritizes issues, risks, and opportunities (Figure 10.3).
- **How it Works:** Define important areas/issues and internally organize these issues in order of importance. Ask stakeholders (contractors, suppliers, clients, and community) to organize the same issues in order of importance.

Use the results to create a Materiality Matrix that will order the issues according to the importance of all stakeholders confronted with the importance that the company has established.

Those issues that are found to be important for the company and the total of other stakeholders (top right of the image) will be further developed into actions to maximize our positive impact and increase the added value provided.

Here you have an example of a materiality matrix in Figure 10.3.

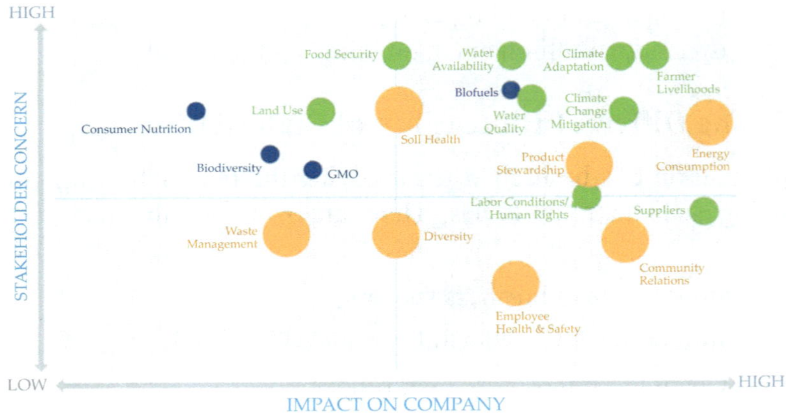

Figure 10.3: Materiality analysis example.

Source: ISO (2017).

10.7.4. Features of Integration

- Integrated thinking, not combined reporting. An integrated report is not about just combining two previously separate publications into one; it is about weaving together a single, integrated narrative.
- A single approach to materiality. An integrated report requires one single threshold for information considered important enough to include.
- An entry point to more detailed information. Rather than pack in every detail, an integrated report should be logically structured with effective navigation and cross-referencing to more information available elsewhere.
- Based on concepts of value creation and capitals. An integrated report explains how the organization interacts with the external environment and six capitals (financial, manufactured, intellectual, human, social, and natural) to create value over the short, medium, and long term.

10.7.5. Pathways to Integrated Reporting

An integrated set of reports-not one report alone meets the needs of diverse report users.

Same look and feel: Have common branding, design, and format.

- **Same Theme:** Highlight the same basic narrative, conveyed in a CEO letter.
- **Same Performance Overview:** Contain one performance dashboard, covering financial and sustainability.
- **Same Approval Process:** Follow identical review and sign-off protocols.

10.7.6. Interactivity

Thinking through the difference between a report and reporting:

- Some stakeholders want a "single moment in time" annual sustainability report, while others expect a depth of analysis that cannot be contained in a printed report. Companies are responding through the creation of a reporting system to satisfy multiple stakeholders and allow for customization.

10.7.7. Leveraging Web 2.0

Companies recognize the dynamic debates and insights are to be found on social media. Some are sharing their sustainability agenda and inviting contributions from stakeholders. Many are translating data into pictures, videos, snippets, blogs, contacts, and posts.

10.7.8. Key Takeaway

We are in the era of disruption (and innovation), where one size does not fit all companies are rethinking reporting approaches and forging new reporting trends which deliver value across multiple dimensions.

CHAPTER 11

CERTIFICATIONS BODIES AND STANDARDS

CONTENTS

11.1. UNITED NATIONS

The UN Global Compact intends to prepare a worldwide development of feasible organizations and partners to make the world they need. That is their vision.

To get this going, the UN Global Compact backings organizations to:

- Do business dependably by adjusting their techniques and operations to 10 Principles on human rights, work, condition, and against defilement; and
- Take key activities to progress more extensive societal objectives, for example, the UN sustainable development goals (SDGs), with an accentuation on cooperation and advancement.

11.2. UNITED NATIONS GLOBAL COMPACT: 10 PRINCIPLES

11.2.1. Human Rights

- **Principle 1:** Businesses should support and respect the protection of internationally proclaimed human rights.
- **Principle 2:** Make sure that they are not complicit in human rights abuses.

11.2.2. Labor

- **Principle 3:** Businesses should uphold the freedom of association and the effective recognition of the right to collective bargaining;
- **Principle 4:** The elimination of all forms of forced and compulsory labor;
- **Principle 5:** The effective abolition of child labor;
- **Principle 6:** The elimination of discrimination in respect of employment and occupation.

11.2.3. Environment

- **Principle 7:** Businesses should support a precautionary approach to environmental challenges;
- **Principle 8:** Undertake initiatives to promote greater environmental responsibility;

- **Principle 9:** Encourage the development and diffusion of environmentally friendly technologies.

11.2.4. Anti-Corruption

- **Principle 10:** Businesses should work against corruption in all its forms, including extortion and bribery.

11.3. RESULTS (FIGURE 11.1)

9,269 COMPANIES	166 COUNTRIES	43,142 PUBLIC REPORTS

Figure 11.1: UNGC certifications figures.

Source: UNGC (2017).

11.3.1. Sustainable Development Goals (SDGs): 17 Goals to Transform Our World

Governments, businesses, and civil society together with the United Nations have started to mobilize efforts to achieve the Sustainable Development Agenda by 2030. Universal, inclusive, and indivisible, the Agenda calls for action by all countries to improve the lives of people everywhere.

In 2015, countries adopted the 2030 Agenda for Sustainable Development and its 17 SDGs. In 2016, the Paris Agreement on climate change entered into force, addressing the need to limit the rise of global temperatures (Figure 11.2).

Figure 11.2: UNGC sustainable development goals.

Source: UNGC (2017).

11.4. ISO 26000

11.4.1. What Advantages Can Be Accomplished by Updating ISO 26000?

An association's execution on social obligation can impact, in addition to other things:

- Competitive preferred standpoint;
- Reputation;
- The capacity to draw in and hold laborers or individuals, clients, customers, and clients;
- The upkeep of worker resolve, duty, and profitability;
- The impression of speculators, proprietors, givers, supports, and the money related group;
- Relationships with organizations, governments, the media, providers, companions, clients, and the group in which it works.

11.4.2. Who Can Profit by ISO 26000 and How?

ISO 26000 gives direction to a wide range of association, paying little respect to their size or area, on:

- Ideas, terms, and definitions identified with a social duty;
- Foundation, patterns, and qualities of social duty;
- Standards and works on identifying with a social duty;
- Center subjects and issues of social duty;
- Coordinating, executing, and advancing socially capable conduct all through the association and, through its approaches and practices, inside its range of prominence;
- Recognizing and drawing in with partners;
- Conveying duties, execution, and other data identified with social obligation.

ISO 26000 is proposed to help businesses in adding to sustainable improvement. It urges them to go past legitimate consistency, perceiving that consistency with the law is a key obligation of any association and a fundamental piece of their social duty program.

The standard tries to advance a typical comprehension of social obligation while supplementing-however, not supplanting other existing instruments and activities. While applying ISO 26000, businesses ought to consider societal, natural, lawful, social, political, and hierarchical assorted qualities and contrast in monetary conditions, while being predictable with global standards of conduct.

Not for confirmation ISO 26000 is not an administration framework standard. It does not contain necessities and, all things considered, cannot be utilized for confirmation. Any offer to affirm, or claims to be ensured, against ISO 26000 would be a deception of its goal and reason.

What specifically does ISO 26000 contain? ISO 26000 addresses seven core subjects of social responsibility defined in the standard and portrayed in the following graphic (Figure 8.3).

Figure 11.3: ISO 26000 framework.

Source: ISO (2017).

11.5. GRI

Global reporting initiative (GRI) that helps organizations, governments, and different businesses comprehend and share the impacts of business on basic sustainability issues, for example, environmental change, human rights, defilement, and numerous others.

Their vision is to make a future where sustainability is essential to each association's basic leadership preparation.

Their central goal is to engage leaders all over the place, through our sustainability models and multi-partner system, to make a move towards a more economical economy and world.

They accept:

- In the energy of a multi-partner handle and comprehensive system;
- Transparency is an impetus for change;
- Our measures engage educated basic leadership;
- A worldwide point of view is expected to change the world;
- Public intrigue ought to drive each choice an association makes.

11.5.1. Vital Needs

GRI is based upon a novel multi-partner rule which guarantees the interest and aptitude of assorted partners in the improvement of its gauges. The advantage of this expansive and comprehensive approach is that businesses can have trust in the vigor of partner cooperation while considering an extensive variety of sustainability issues. GRI is seeing a more prominent interest in sustainability data for particular uses exhibited by the development of new reporting activities. Although the utilization of the data varies, the inspiration for giving it is regularly the same: to assemble trust. (Figure 11.4).

Figure 11.4: GRI framework.

Source: GRI (2017).

11.5.2. Empowering Smart Policy

GRI's worldwide group traverses a large number of columnists, yet our degree stretches out past topography and into different fields of work. They work with governments, worldwide businesses, and capital markets. Thus, in this work, 27 nations utilize GRI in their sustainability approaches and seek us for direction as the universes most broadly utilized sustainability detailing benchmarks. Furthermore, they have long-standing joint efforts with more than 20 universal businesses, for example, the UNGC, OECD, and the UN Working Group on Business and Human Rights.

GRI teams up with different businesses on various worldwide driving edge activities which concentrate on handling a portion of the world's most squeezing difficulties, for example, environmental change, human rights, and defilement. Through this sort of joint effort, they are finding imaginative methods for utilizing sustainability detailing data to help bolster organizations specifically.

Consistently they help developed nations put millions in sustainable improvement exercises in developing markets to address neediness-related issues, for example, water utilizes, arrive utilize and gender equality correspondence. GRI's Standards keep on playing an imperative part in these sorts of speculation streams guaranteeing that sustainability issues are mulled over by these businesses as they improve their aggressive edge in the neighborhood and global markets.

11.5.3. More Reporters, Better Reporting

GRI is best known for driving the improvement of the sustainability transparency procedure and keeps on building their worldwide group which incorporates a great many prepared professionals and columnists. Among this differing group, they additionally have many center individuals who work straightforwardly with them to champion their central goal.

As a worldwide autonomous association, the provincial centers empower them to address novel issues crosswise over topographies and set up a genuinely worldwide nearness. GRI works with a developing group to create industry-driving exercises to promote the estimation of sustainability detailing. They support and empower those businesses that do not right now answer to do as such, particularly advancing the take-up of transparency among small and medium-sized ventures.

GRI persuades and underpins those that as of now do answer to make progress toward a superior nature of transparency through our preparation and information sharing exercises. What is more, for those that are driving the way, they work in a coordinated effort to research and imagine enormous picture difficulties, for example, the eventual fate of transparency through their corporate initiative engagement exercises.

GRI is focused on giving worldwide models to sustainability detailing and keeps on strengthening their believability and part as a standard-setter through late changes to administration structure. This incorporates the production of a different administration body to create and support GRI's Sustainability Reporting Standards to expand the autonomy of these exercises.

Guaranteeing that more organizations and governments' esteem sustainability data will remain an essential piece of GRIs DNA, however, the time has desired them to expand on this inheritance and catalyze another time where organizations and governments settle on vital choices in light of these vital sustainability issues.

11.5.4. Moving Beyond Reports

GRI's concentrate has dependably been on the detailing procedure and the estimation of the data that originates from it. While the sustainability report remains a vital yield of the detailing procedure, they should now move past the report itself to guarantee that chiefs have entry to the high caliber and solid data they are progressively requesting.

From the utilization of sustainability data in business insight to the real formation of new lines of business, unmistakably, this data is not only for sustainability experts. It is for financial specialists to empower more prominent trust in their ventures, clients in their products and administrations, subjects in their legislatures, et cetera.

These gatherings and people all need distinctive data and need to get to it in various ways. Today we are seeing a rush of new routes in which data from the sustainability detailing procedure is being utilized by differing gatherings of partners to address their particular needs.

Yet, for this data to genuinely enable economical choices in each association, it must be more open, equivalent, and accessible progressively. GRIs part in deduction past the sustainability report is to help the inexorably various scope of clients take advantage of the sustainability data they require, empowering them to utilize this data in more imaginative ways.

11.5.5. Advancement and Collaboration

The sustainability scene is continually changing and GRI has dependably trusted in the requirement for development. GRI's Standards are a free, open great which have effectively motivated the making of new systems, businesses, arrangements, and professions all centered around enhancing the estimation of sustainability data.

GRI keeps on driving the exchange on incorporating sustainability and monetary data, working with various businesses and partners to further this plan. Effective development requires another soul of joint effort and comprehensiveness in the reporting field, guaranteeing businesses cooperate towards a shared objective of building trust in organizations and governments. A portion of the best developments, be that as it may, will originate from coordinated efforts past the detailing field.

This is the place innovation and enormous information have a gigantic part to play and we are working with the world's driving innovation organizations to move to future 'past reports.' This will help all partners get to the data they have to settle on better business and approach choices, in light of the world's most generally utilized and trusted sustainability detailing benchmarks.

11.5.6. Global Reporting Initiative (GRI)-G4

Drivers:

- Increasing understanding of the connection between ESG issues and sustainable business value;
- Increasing interest in organizational focus-what and where.

BIBLIOGRAPHY

1. Adams, A. V., (2009). Skills development in the informal sector of sub-Saharan Africa. In: Maclean, R., & Wilson, D., (eds.), *International Handbook of Education for the Changing World of Work: Bridging Academic and Vocational Education, UNESCO-UNEVOC* (Vol. 6, pp. cxiii–cxxxiii). New York, Springer.
2. Adams, A. V., Razmara, S., & Johansson, D. S. S., (2013). *Improving Skills Development in the Informal Sector: Strategies for Sub-Saharan Africa.* Washington, DC, World Bank.
3. ADEA, (2012). *Promoting Critical Skills for the Accelerated and Sustainable Development of Africa: The Key Messages of the ADEA Triennale*. Ouagadougou, Burkina Faso.
4. African Union, (2007). *Strategy to Revitalize Technical and Vocational Education and Training in Africa*. Addis Ababa, African Union. https://www.academia.edu/25379014/Strategy_to_Revitalize_Technical_and_Vocational_Education_and_Training_TVET_in_Africa_Final_Draft (accessed on 18 June 2022).
5. Aguinis, H., (2004). *Regression Analysis for Categorical Moderators*. New York: Guilford.
6. Aguinis, H., (2009). *Performance Management* (2nd edn.). Upper Saddle River, NJ: Pearson Prentice Hall.
7. Aguinis, H., & Branstetter, S. A., (2007). Teaching the concept of the sampling distribution of the mean. *J. Manag. Educ., 31*, 467–483.

8. Aguinis, H., & Pierce, C. A., (2008). Enhancing the relevance of organizational behavior by embracing performance management research. *J. Organ. Behav., 29*, 139–145.
9. Aguinis, H., Beaty, J. C., Boik, R. J., & Pierce, C. A., (2005). Effect size and power in assessing moderating effects of categorical variables using multiple regression: A 30-year review. *J. Appl. Psychol., 90*, 94–107.
10. Aguinis, H., Mazurkiewicz, M. D., & Heggestad, E. D., (2009). Using Web-based frame-of-reference training to decrease biases in personality-based job analysis: An experimental field study. *Pers. Psychol.* In press.
11. Aguinis, H., Sturman, M. C., & Pierce, C. A., (2008). Comparison of three meta-analytic procedures for estimating moderating effects of categorical variables. *Organ. Res. Methods, 11*, 9–34.
12. Akoojee, S., (2011). Private TVET provision in Africa. *Preserving Traditional Forms Differently- Managing Novel Ones Creatively: Understanding Current Private Provision in Africa and Policy Implications*. Background paper commissioned by UNESCO Education Sector.
13. Allais, S., (2010). *The Implementation and Impact of National Qualifications Frameworks: Report of a Study in 16 Countries*. Geneva, Skills and Employability Department, ILO.
14. Alliger, G. M., Tannenbaum, S. I., Bennett, W. Jr., Traver, H., & Shotland, A., (1997). A meta-analysis of the relations among training criteria. *Pers. Psychol., 50*, 341–358.
15. Anda, M. L. D., (2009). *Qualifications Frameworks: Implementation and Impact: Background Case Study on Mexico.* Geneva, Skills and Employability Department, ILO.
16. Aragon, S., ´ anchez, A., Barba-Arag ´ on, I., & Sanz-Valle, R., (2003). Effects of training on business results. ´ *Int. J. Hum. Resour. Manag., 14*, 956–980.
17. Arandarenko, M., & Bartlett, W., (2012). *Labor Market and Skills in the Western Balkans*, 60–70. Belgrade, FREN.
18. Arthur, W. J., Bennett, W. J., Edens, P., & Bell, S. T., (2003). Effectiveness of training in organizations: A meta-analysis of design and evaluation features. *J. Appl. Psychol., 88*, 234–245.

19. ASEAN, (2009). *Cha-Am Hua Hin Declaration on Strengthening Cooperation on Education to Achieve an ASEAN Caring and Sharing Community*, 17–52.

20. Asian Development Bank, (2009). *Education and Skills: Strategies for Accelerated Development in Asia and the Pacific.* Manila, Asian Development Bank.

21. Bacchetta, M., & Jansen, M., (2011). *Making Globalization Socially Sustainable.* Geneva, ILO and WTO.

22. Baethge, M., Achtenhagen, F., Arends, L., Babic, E., Baethge-Kinsky, V., & Weber, S., (2006). *PISA- VET: A Feasibility-Study.* Stuttgart, Germany, Steiner.

23. Baranzini, D., Bacchi, M., & Cacciabue, P. C., (2001). A tool for evaluation and identification of training needs in aircraft. *Hum. Factors Aerosp. Saf., 1*, 167–193.

24. Barber, J., (2004). Skill upgrading within informal training: Lessons from the Indian auto mechanic. *Int. J. Train. Dev., 8,* 128–139.

25. Bartels, F. B., & Lederer, S., (2009). *Outsourcing Markets in Services: International Business Trends,* 20-35.

26. Bartlett, K. R., & Rodgers, J., (2004). HRD as national policy in the Pacific Islands. *Adv. Dev. Hum. Resour., 6,* 307–314.

27. Basu, C. K., & Majumdar, S., (2010). The role of ICT and TVET in rural development and poverty alleviation. *International Handbook of Education for the Changing World of Work* (Vol. 1, pp. 1923–1934). UNESCO-UNEVOC. Dordrecht, Springer.

28. Becker, G., (1962). Investment in human capital: A theoretical analysis. *J. Pol. Econ., 70*(5, Pt. 2), S9–49.

29. Becker, G., (1964). *Human Capital: A Theoretical and Empirical Analysis, With Special Reference to Education.* New York: Columbia Univ. Press.

30. Bell, B. S., & Kozlowski, S. W. J., (2002). Adaptive guidance: Enhancing self-regulation, knowledge, and performance in technology-based training. *Pers. Psychol., 55*, 267–306.

31. Ben, S. H., (2009). Études De Cas: La tunisie. *Enseignement Supérieur Et Marché Du Travail Dans Le Monde Arabe.* Beyrouth, Institut français du Proche-Orient.

32. Benson, G., Finegold, D., & Mohrman, S. A., (2004). You paid for the skills, now keep them: Tuition reimbursement and voluntary turnover. *Acad. Manag. J., 47*, 315–331.

33. Besley, T., & Cord, L., (2006). *Delivering on the Promise of Pro-Poor Growth Insights and Lessons from Country Experiences*. Basingstoke, UK/New York/ Washington, DC, Palgrave Macmillan, and World Bank.

34. Bewick, T., & Abbott, P., (2010). *Think Global, Act Sectoral.* London, International Network of Sectoral Skills Organizations.

35. Bhawuk, D. P. S., & Brislin, R. W., (2000). Cross-cultural training: A review. *Appl. Psychol.: An Int. Rev., 49*, 162–191.

36. Bhorat, H., & Lundall, P., (2002). *Employment, Wages and Skills Development: Firm-Specific Effects-Evidence from Two Firm Surveys in South Africa, Working Paper No. 02/68*. Cape Town, South Africa, University of Cape Town Development Policy Research Unit.
37. Bhorat, H., Lundall, P., & Rospabe, S., (2002). *The South African Labor Market in a Globalizing World: Economic and Legislative Considerations*. Geneva, ILO.
38. Biggs, T., Miller, M., Otto, C., & Tyler, G., (1996). *Africa Can Compete! Export Opportunities and Challenges for Garments and Home Products in the European Market*. Washington, DC, World Bank.
39. Blake, D., & Gallagher, D., (2009). Examining the development of the Victorian certificate of applied learning and its implications in schools and teacher education in Australia. *Journal of Applied Learning in Higher Education* (Vol. 1, pp. 49–71). Missouri Western State University.
40. Blanchard, P. N., & Thacker, J. W., (2007). *Effective Training: Systems, Strategies, and Practices* (3rd edn.). Upper Saddle River, NJ: Pearson Prentice Hall.
41. BMBF, (2015). *Reform of Vocational Education and Training*, 23–68.
42. Bozell, M., & Liston, C., (2010). *Building Effective Green Energy Programs in Community Colleges*. New York, Workforce Strategy Center.
43. Brown, K. G., (2001). Using computers to deliver training: Which employees learn and why? *Pers. Psychol., 54,* 271–296.
44. Brown, K. G., (2005). Examining the structure and nomological network of trainee reactions: A closer look at "smile sheets." *J. Appl. Psychol., 90*, 991–1001.
45. Brown, K. G., & Van, B. M. E., (2007). Applying a social capital perspective to the evaluation of distance training. In: Fiore, S. M., & Salas, E., (eds.), *Where is the Learning in Distance Learning? Towards a Science of Distributed Learning and Training* (pp. 41–63). Washington, DC: Am. Psychol. Assoc.
46. Brown, M., (2012). Developing green skills through a regional TAFE institute. *Paper Presented at the 15th AVETRA Conference* (pp. 38–61). Canberra.
47. Brown, P., Green, A., & Lauder, H., (2001). *High Skills: Globalization, Competitiveness, and Skill Formation*. Oxford, Oxford University Press.

48. Brown, P., Lauder, H., & Ashton, D., (2011). *The Global Auction: The Broken Promises of Education, Jobs, and Incomes*. Oxford, Oxford University Press.

49. Brown, T. C., (2005). Effectiveness of distal and proximal goals as transfer-of-training interventions: A field experiment. *Hum. Resour. Dev. Q., 16*, 369–387.

50. Campbell, J. P., (1971). Personnel training and development. *Annu. Rev. Psychol., 22*, 565–602.

51. CANTA, (2013). *Regional TVET Strategy for Workforce Development and Economic Competitiveness* (pp. 17–52). Port of Spain, CANTA.

52. Cantelmo, A., Giovanni, M., & Chris, P., (2019). *Macroeconomic Outcomes in Disaster-Prone Countries*. International Monetary Fund Working Paper 19/217. https://www.imf.org/en/Publications/WP/Issues/2019/10/11/Macroeconomic-Outcomes-in-Disaster-Prone-Countries-48704 (accessed on 18 June 2022).

53. Cascio, W. F., & Aguinis, H., (2005). *Applied Psychology in Human Resource Management* (6th edn.). Upper Saddle River, NJ: Pearson Prentice Hall.

54. Cascio, W. F., & Aguinis, H., (2008). Staffing twenty-first-century organizations. In: Walsh, J. P., & Brief, A. P., (eds.), *Academy of Management Annals* (Vol. 2, pp. 133–165). Mahwah, NJ: Erlbaum.

55. Castejon, J. M., Coles, M., Chakroun, B., Deij, A., & McBride, V., (2011). *Developing Qualifications Frameworks in EU Partner Countries: Modernizing Education and Training*. London, Anthem Press.

56. CEDEFOP and European Commission, (2010). *European Inventory on Validation of Non-Formal and Informal Learning,* 14–25.

57. CEDEFOP, (2008). *Social Partners and Sectoral Training Funds: Mobilizing Resources*. Briefing note 3/2008. http://www.cedefop.europa.eu/EN/Files/8560_en.pdf (accessed on 18 June 2022).

58. Chakroun, B., (2008). What can we learn from policy learning? In: Chakroun, B., & Sahlberg, P., (eds.), *ETF Yearbook 2008: Policy Learning in Action* (pp. 11–18). Luxembourg, Office for Official Publications of the European Communities.

59. Chakroun, B., & Jimeno, S. E., (2010). TVET reforms in the Arab region: The 'push' and 'pull' in policy development. In: Mazawi, A.

E., & Sultana, R. G., (eds.), *Education and the Arab 'World': Political Projects, Struggles, and Geometries of Power* (pp. 59–76). New York, Routledge.

60. Chakroun, B., & Sahlberg, P., (2008). *ETF Yearbook 2008: Policy Learning in Action*. Luxembourg, Office for Official Publications of the European Communities.
61. Chen, G., Thomas, B., & Wallace, J. C., (2005). A multilevel examination of the relationships among training outcomes, mediating regulatory processes, and adaptive performance. *J. Appl. Psychol., 90*, 827–841.
62. Cheng, I. H., (2010). Case studies of integrated pedagogy in vocational education: A three-tier approach to empowering vulnerable youth in urban Cambodia. *International Journal of Educational Development, 30*(4), 438–446.
63. Cheng, K., & Marope, P. T. M., (2010). 21st-century development challenges and opportunities: Demands on education systems. Developing knowledge, skills, competencies, and values for the 21st century. In: *14th UNESCO-APEID International Conference*. Plenary Session I. Bangkok, UNESCO Bangkok.
64. Chiaburu, D. S., & Marinova, S. V., (2005). What predicts skill transfer? An exploratory study of goal orientation, training self-efficacy, and organizational supports. *Int. J. Train. Dev., 9*, 110–123.
65. Chinien, C., (2005). *The Use of ICTs in Technical and Vocational Education and Training: South-Eastern Europe: Analytical Survey*. Moscow, UNESCO Institute for Information Technologies in Education.
66. Chinien, C., Boutin, F., McDonald, C., & Searcy, C., (2004). *Skills to Last: Broadly Transferable Sustainable Development Literacy for the Canadian Workforce*. Ottawa, National Center for Workshop Development.
67. Chitolie-Joseph, E., (2008). T*he Development and State of the Art of Adult Learning and Education (ALE): National Report of Saint Lucia*. Castries, Ministry of Education and Culture.
68. Cho, E., & McLean, G. N., (2004). What we discovered about NHRD and what it means for HRD. *Adv. Dev. Hum. Resour., 6*, 382–393.
69. Cho, J., (2012). *Innovation of National Education Administration System in Korea*. Presentation at KRIVET-Africa TVET Experts Workshop.

70. Clardy, A., (2005). Reputation, goodwill, and loss: Entering the employee training review equation. *Hum. Resour. Dev. Rev., 4,* 279–304.

71. Collett, K., (2008). *Vocational Training for Excluded Groups: Success Stories*. London, Centre for Skills Development, City and Guilds.

72. Collins, D. B., & Holton, E. F. III., (2004). The effectiveness of managerial leadership development programs: A meta-analysis of studies from 1982 to 2001. *Hum. Resour. Dev. Q., 15*, 217–248.

73. Colquitt, J. A., LePine, J. A., & Noe, R. A., (2000). Toward an integrative theory of training motivation: A meta-analytic path analysis of 20 years of research. *J. Appl. Psychol., 85*, 678–707.

74. COMEDAF, (2007). Strategy to revitalize technical and vocational education and training (TVET) in Africa. *Presented at the Meeting of the Bureau of the Conference of Ministers of Education of the African Union (COMEDAF II+)*. Addis Ababa.

75. Condon, L., & Rickard, A., (2009). Australia: The role of partnerships, industry skills councils and training packages. In: Fien, J., Maclean, R., & Park, M. G., (eds.), *Work, Learning and Sustainable Development* (pp. 279–294). London, Springer.

76. Council of Australian Governments, (2009). *Green Skills Agreement*. Australian Government, Department of Education, Employment and Workplace Relations.

77. Cullen, J., & Turnbull, S., (2005). A meta-review of the management development literature. *Hum. Resour. Dev. Rev., 4,* 335–355.

78. Cunha, F., & Heckman, J., (2007). The technology of skill formation. *American Economic Review, 97*(2), 31–47.

79. Darch, H., & Lucas, T., (2002). Training as an e-commerce enabler. *J. Workplace Learn.,14,* 148–155.

80. Davis, F. D., & Yi, M. Y., (2004). Improving computer skill training: Behavior modeling, symbolic mental rehearsal, and the role of knowledge structures. *J. Appl. Psychol., 89*, 509–523.

81. Day, D. V., (2000). Leadership development: A review in context. *Leadersh. Q., 11*, 581–614.

82. Decker, P. J., & Nathan, B. R., (1985). *Behavior Modeling Training: Principles and Applications*. New York: Praeger.

83. *Demobilization & Reintegration' Programs* (pp. 50–57). Geneva, Centre d'études et de recherches internationales, Sciences Po, Graduate Institute of International Studies.
84. Department for Education and Skills, (2003). *Aiming High: Raising the Achievement of Gypsy Traveler Pupils: A Guide to Good Practice.* London, Department for Education and Skills.
85. Derksen, D., (2014). *Reintegrating Armed Groups in Afghanistan: Lessons from the Past*. United States Institute of Peace (USIP) Brief 168.Washington, DC, USIP.
86. DeRouin, R. E., Fritzsche, B. A., & Salas, E., (2004). Optimizing e-learning: Research-based guidelines for student-controlled training. *Hum. Resour. Manag. Rev., 43*, 147–162.
87. Driskell, Hall, J. J., & Salas, E., (2001). Does stress training generalize to novel settings? *Hum. Factors, 42*, 99–110.
88. Dvir, T., Eden, D., Avolio, B. J., & Shamir, B., (2002). Impact of transformational leadership on follower development and performance: A field experiment. *Acad. Manag. J., 45,* 735–744.
89. ECOTEC, (2008). *A European Inventory on Validation of Non-Formal and Informal Learning*. Birmingham, ECOTEC Research and Consulting.
90. Edkins, G. D., (2002). A review of the benefits of aviation human factors training. *Hum. Factors Aerosp. Saf., 2,* 201–216.
91. Education Development Centre, (2011). *The Maldivian National Curriculum Framework (Fifth Working Draft)*. Malé, Ministry of Education of the Maldives.
92. Eichhorst, W., Escudero, V., Marx, P., & Tobin, S., (2010). *The Impact of the Crisis on Employment and the Role of Labor Market Institutions.* Discussion Paper 5320. Bonn, Germany, IZA. http://ftp.iza.org/dp5320.pdf (accessed on 18 June 2022).
93. Ellis, A. P. J., Bell, B. S., Ployhart, R. E., Hollenbeck, J. R., & Ilgen, D. R., (2005). An evaluation of generic teamwork skills training with action teams: Effects on cognitive and skill-based outcomes. *Pers. Psychol., 58*, 641–672.
94. Employers' Forum on Disability, (2009). *What Does the Research Say are the Commercial Benefits?,* 38–52.

95. ETF and World Bank, (2005). *Reforming Technical Vocational Education and Training in the Middle East and North Africa: Experiences and Challenges*. Luxembourg, Office for Official Publications of the European Communities.
96. ETF, (2008). *Peer Review Report: National Qualifications Framework (NQF) Development in Egypt*, 48–51.
97. ETVET Council, & Leney, T., (2009). *CAQA Organization: Rationale and Brief Functional Description*. Unpublished paper.
98. EU High-Level Group, (2012). *EU High-Level Group of Experts on Literacy: Final Report*. Luxembourg, European Commission.
99. Eurofound, (2010). *European Company Survey 2009*. Dublin, Euro found.
100. European Alliance for Apprenticeships, (2013). *Work-Based Learning in Europe: Renewing Traditions: Background Information*. Bonn, International BIBB Conference 2013.
101. European Commission, (1999). *Convergences and Divergences in European Education and Training Systems*. Education, Training, Youth research project commissioned by the ECDG22.
102. Fares, J., & Puerto, S., (2009). *Towards Comprehensive Training*. Social protection, Discussion paper no. SP 0924. Washington, DC, World Bank. http://documents.worldbank.org/curated/en/2009/11/11510314/towards-comprehensive-training (accessed on 18 June 2022).
103. Faust, M., Voskamp, U., & Wittke, V., (2004). *Globalization and the Future of National Systems: Exploring Patterns of Industrial Reorganization and Relocation in an Enlarged Europe* (pp. 14–72). Gottingen, Organizations Beratung Importgut Exportschlager (OBIE).
104. Fien, J., (2013). *Education for Sustainable Development in Technical and Vocational Education and Workplace Training*. Background paper prepared for Decade for Education for Sustainable Development Final Report.
105. Fien, J., & Maclean, R., (2009). Introduction: The legacy of the Bonn declaration. In: Fien, J., Maclean, R., & Park, M. G., (eds.), *Work, Learning and Sustainable Development* (p. xix). Dordrecht, Springer.
106. Fien, J., & Wilson, D., (2005). *Promoting Sustainable Development in TVET: The Bonn Declaration, Prospects, 35*(3), 273–288.

107. Fiol, C. M., O'Connor, E. J., & Aguinis, H., (2001). All for one and one for all? The development and transfer of power across organizational levels. *Acad. Manag. Rev., 26*, 224–242.

108. Ford, J. K., & Kraiger, K., (1995). The application of cognitive constructs and principles to the instructional systems model of training. Implications for needs assessment, design, and transfer. In: Cooper, C. L., & Robertson, I. T., (eds.), *International Review of Industrial and Organizational Psychology* (Vol. 10, pp. 1–48). Chichester, UK: Wiley.

109. Ford, J. K., & Schmidt, A. M., (2000). Emergency response training: Strategies for enhancing real-world performance. *J. Hazard. Mater., 75*, 195–215.

110. Fowlkes, J. E., Salas, E., Baker, D. P., Cannon-Bowers, J. A., & Stout, R. J., (2000). The utility of event-based knowledge elicitation. *Hum. Factors, 42*, 24–35.

111. Fox, A., (2003). Training budgets are said to be withstanding companies' economic troubles. *HR Mag., 48*(7), 32, 38.

112. Frayne, C. A., & Geringer, J. M., (2000). Self-management training for improving job performance: A field experiment involving salespeople. *J. Appl. Psychol., 85*, 361–372.

113. Frese, M., Beimel, S., & Schoenborn, S., (2003). Action training for charismatic leadership: Two evaluations of studies of a commercial training module on inspirational communication of a vision. *Pers. Psychol., 56*, 671–697.

114. G20, (2013). *G20 Leaders' Declaration*. Saint Petersburg.

115. Gajaweera, G. A. K., (2010). *Qualifications Frameworks: Implementation and Impact Background.* Case Study on Sri Lanka. Geneva, Skills and Employability Department, ILO.

116. Gasskov, V., (2006a). *Mechanisms of Financing TVET* (pp. 128–132). Presentation for GTZ symposium, Addis-Ababa.

117. Gaudine, A. P., & Saks, A. M., (2004). A longitudinal quasi-experiment on the effects of post-training transfer interventions. *Hum. Resour. Dev. Q., 15*, 57–76.

118. Gereffi, G., (2005). *The New Off-Shoring of Jobs and Global Development* (pp. 187–190). Geneva, IILS.

119. Global Commission on International Migration, (2005). *Migration in an Interconnected World: New Directions for Action*. Geneva, ILO.

120. Goeters, K. M., (2002). Evaluation of the effects of CRM training by the assessment of non-technical skills under LOFT. *Hum. Factors Aerosp. Saf., 2*, 71–86.

121. Goldstein, I. L., (1980). Training in work organizations. *Annu. Rev. Psychol., 31*, 229–272.

122. Goldstein, I. L., & Ford, J. K., (2002). *Training in Organizations* (4th edn.). Belmont, CA: Wadsworth.

123. Government of Mauritius, (2001). *Mauritius Qualifications Authority Bill 2001, No. XXXIX*, 11, 12.

124. Government of New South Wales, (2013). *Department of Environment, Climate Change and Water.* website. https://www.environment.nsw.gov.au (accessed on 18 June 2022).

125. Graziosi, A., (2011). *Training Institutions and Labor Market Needs: Bridging the Gap.* Geneva, ILO.

126. Grootings, P., & Nielsen, S., (2008). Policy learning: AppG Psacharopoulos

127. lying the changing learning paradigm for policy advice on TVET reforms in transition countries. In: CEDEFOP, (ed.), *Modernizing Vocational Education and Training* (Vol. 2). Fourth report on vocational education and training research in Europe: Background report. Luxembourg: Office for Official Publications of the European Communities.

128. GTZ, (2006). *Implementation Issues of Diversified Financing Strategies for TVET* (pp. 21–63). Eschborn, GTZ.

129. Guerrero, S., & Barraud-Didier, V., (2004). High-involvement practices and performance of French firms. *Int. J. Hum. Resour. Manag., 15*, 1408–1423.

130. Hakim, G. L., (2011). *Does TVET Promote Social Mobility?* Background paper commissioned by UNESCO Education Sector.

131. Hammonds, K. H., (2005). *Why we Hate HR* (Vol. 97, p. 40). Fast Co.

132. Handel, J. M., (2008). *Measuring Job Content: Skills, Technology, and Management Practices*. Institute for Research for Poverty Discussion Paper No. 1357–08.

133. Hanushek, E., Woessmann, L., & Zhang, L., (2011). *General Education, Vocational Education and Labor-Market Outcomes Over the Life-Cycle*. Bonn, Institute for the Study of Labor.

134. Hargroves, K., & Smith, M. H., (2005). *The Natural Advantage of Nations: Business Opportunities, Innovation and Governance in the 21st Century* (pp. 160–170). London, Earthscan.

135. Härtl, M., (2009). Germany: Policymaking strategies and project experiences. In: Fien, J., Maclean, R., & Park, M. G., (eds.), *Work, Learning and Sustainable Development*. Dordrecht, Springer.

136. Heart Trust, N. T. A., (2009). *Policy on Competency-Based Education and Training*. Draft Concept Paper. Kingston, HEART Trust NTA.

137. Heimbeck, D., Frese, M., Sonnentag, S., & Keith, N., (2003). Integrating errors into the training process: The function of error management instructions and the role of goal orientation. *Pers. Psychol., 56*, 333–361.

138. Hill, C. E., & Lent, R. W., (2006). A narrative and meta-analytic review of helping skills training: Time to revive a dormant area of inquiry. *Psychother. Theory Res. Pract., 43*, 154–172.

139. Hiromichi, K., (2011). *Trends in Technical and Vocational Education and Training*. Background paper commissioned by UNESCO Education Sector.

140. Holladay, C. L., & Quinones, M. A., (2003). Practice variability and transfer of training: The role of self-efficacy generality. *J. Appl. Psychol., 88*, 1094–1103.

141. Holmes, K., (2009). The reform and governance of public TVET institutions: Comparative experiences. In: Maclean, R., & Wilson, D., (eds.), *International Handbook of Education for the Changing World of Work: Bridging Academic and Vocational Learning* (pp. 905–919). Dordrecht, Springer.

142. Holton, E. F. III., (2003). Cycle time: A missing dimension in HRD research and theory. *Hum. Resour. Dev. Rev., 2,* 335, 336.

143. Holton, E. F. III., (2005). Holton's evaluation model: New evidence and construct elaborations. *Adv. Dev. Hum. Resour., 7,* 37–54.

144. Holton, E. F. III., Bates, R. A., & Ruona, W. E. A., (2001). Development of a generalized learning transfer system inventory. *Hum. Resour. Dev. Q., 11*, 333–360.

145. Holton, E. F. III., Chen, H. C., & Naquin, S. S., (2003). An examination of learning transfer system characteristics across organizational settings. *Hum. Resour. Dev. Q., 14*, 459–482.

146. Huint, H., & Saks, A. M., (2003). Translating training science into practice: A study of managers' reactions to post-training transfer interventions. *Hum. Resour. Dev. Q., 14*, 181–198.

147. Ibarrarán, P., & Rosas-Shady, D., (2008). *Evaluating the Impact of Job Training Programs in Latin America: Evidence from IDB Funded Operations*. Washington, DC, IADB.

148. ICRC, (2020). *When Rain Turns to Dust: Understanding and Responding to the Combined Impact of Armed Conflicts and the Climate and Environment Crisis on People's Lives*. https://www.icrc.org/sites/default/files/topic/file_plus_list/rain_turns_to_dust_climate_change_conflict.pdf (accessed on 18 June 2022).

149. ILO and CEDEFOP, (2011). Skills for green jobs. *A Global View: Synthesis Report Based on 21 Country Studies.* Geneva, ILO. http://www.ilo.org/wcmsp5/groups/public/---dgreports/---dcomm/---publ/documents/publication/wcms_159585.pdf (accessed on 18 June 2022).

150. ILO and IILS, (2011). *World of Work Report 2011: Making Markets Work for Jobs*. Geneva, ILO.

151. ILO, (2006). Changing patterns in the world of work, report of the director-general. *International Labor Conference, 95th Session*. Geneva, ILO.

152. Inter-Agency Working Group on TVET Indicators, (2012). *Proposed Indicators for Assessing Technical and Vocational Education and Training.* Working document.

153. International Fund for Agricultural Development, (2010). *Rural Poverty Report 2011: Enabling Poor Rural People to Overcome Poverty.* Rome, IFAD.

154. International Monetary Fund, (2019). *Fiscal Monitor, October 2019: Mitigating Climate Change.* https://www.imf.org/en/Publications/FM/Issues/2019/09/12/fiscal-monitor-october-2019 (accessed on 18 June 2022).

155. International Monetary Fund, (2020a). *Sub-Saharan Africa Regional Economic Outlook,* April 2020. COVID-19: An unprecedented threat to development, chapter 2, Adapting to climate change in Sub-Saharan Africa (SSA), https://www.imf.org/en/Publications/REO/SSA/Issues/2020/04/01/sreo0420 (accessed on 18 June 2022).

156. International Monetary Fund, (2020b). *World Economic Outlook*, October 2020. A long and difficult ascent, chapter 3, mitigating climate change—growth- and distribution-friendly strategies, (accessed on 18 June 2022).
157. International Monetary Fund, 2017: World Economic Outlook, (2017). *Seeking Sustainable Growth: Short-Term Recovery, Long-Term Challenges.* Chapter 3, The Effects of Weather Shocks on Economic Activity: How Can Low-Income Countries Cope? https://www.imf.org/en/Publications/WEO/Issues/2017/09/19/world-economic-outlook-october-2017 (accessed on 18 June 2022).
158. International Organization for Migration, (2010). *World Migration Report 2010: The Future of Migration: Building Capacities for Change.* Geneva, International Organization for Migration.
159. International Skills Standards Organization, (2010). *Think Global Act Sectoral* (pp. 15–19). London, INSSO.
160. IPCC, (2014). Climate change 2014: Synthesis report. In: Core Writing Team, Pachauri, R. K., & Meyer, L. A., (eds.), *Contribution of Working Groups I, II, and III to the Fifth Assessment Report of the Intergovernmental Panel on Climate Change* (p. 151). IPCC, Geneva, Switzerland. https://www.ipcc.ch/report/ar5/syr/ (accessed on 18 June 2022).
161. Jacinto, C., (2011). *Technical and Vocational Education and Training in the Last Decade in Latin America.* Background paper commissioned by UNESCO Education Sector.
162. Jacobsen, J., Lindekaer, A. L., Ostergaard, H. T., Nielsen, K., Ostergaard, D., Laub, M., et al., (2001). Management of anaphylactic shock evaluated using a full-scale anesthesia simulator. *Acta Anaesthesiol. Scand., 45,* 315–319.
163. Jacoby, S. M., (2004). *Employing Bureaucracy* (2nd edn.). Mahwah, NJ: Erlbaum.
164. Janjua, S., (2011). Is skill training a good investment for the poor? The evidence from Pakistan. *International Journal of Training Research, 9*, 95–109.
165. Jaramillo, A., & Angel-Urdinola, D., (2011). Higher Education, productivity, and labor market insertion: Are MENA countries getting results? In: Jaramillo, A., & Melonio, T., (eds.), *Breaking Even or Breaking Through: Reaching Financial Sustainability While Providing High Quality Standards in Higher Education in the Middle East and North Africa* (pp. 1–12). Washington, DC, World Bank.

166. Johanson, R. K., & Adams, A. V., (2004). *Skills Development in Sub-Saharan Africa*. Washington, DC, World Bank.

167. Johanson, R., (2009). *A Review of National Training Funds, SP Discussion Paper No. 0922*. Washington, DC, World Bank Social Protection Unit.

168. Kaufman, R., & Guerra, I., (2001). A perspective adjustment to add value to external clients (including society). *Hum. Resour. Dev. Q., 12*, 319–324.

169. Kearns, P., (2009). What are the limits of ICTs and media in the delivery of TVET? An Australian focus. In: Maclean, R., & Wilson, D., (eds.), *International Handbook of Education for the Changing World of Work*. Dordrecht, Springer.

170. Keevy, J., (2011). *The Recognition of Qualifications Across Borders: The Contribution of Regional Qualifications Frameworks*. Background paper commissioned by UNESCO. Pretoria, South African Qualifications Authority.

171. Keevy, J., Chakroun, B., & Deij, A., (2010). *Transnational Qualifications Frameworks*. Luxembourg, Publications Office of the European Union.

172. Keith, N., & Frese, M., (2005). Self-regulation in error management training: Emotion control and metacognition as mediators of performance effects. *J. Appl. Psychol., 90*, 677–691.

173. Keith, N., & Frese, M., (2008). Effectiveness of error management training: A meta-analysis. *J. Appl. Psychol., 93*, 59–69.

174. Khan, S., (1999). *Towards Sustainability: A Guide for Colleges*. London, Further Education Development Agency.

175. King, K., & McGrath, S., (2002). *Globalization, Knowledge and Enterprise*. Oxford, Symposium.

176. King, K., & Palmer, R., (2010). *Planning for Technical and Vocational Skills Development*. Fundamentals of Educational Planning, No. 94. Paris, UNESCO International Institute for Educational Planning.

177. Klein, H. J., Noe, R. A., & Wang, C., (2006). Motivation to learn and course outcomes: The impact of delivery mode, learning goal orientation, and perceived barriers and enablers. *Pers. Psychol., 59*, 665–702.

178. Kontoghiorghes, C., (2004). Reconceptualizing the learning transfer conceptual framework: Empirical validation of a new systemic model. *Int. J. Train. Dev., 8*, 210–221.

179. Kozlowski, S. W. J., & Salas, E., (1997). An organizational systems approach for the implementation and transfer of training. In: Ford, K., Kozlowski, S. W., Kraiger, K., Salas, E., & Teachout, M., (eds.), *Improving Training Effectiveness in Work Organizations* (pp. 247–287). Mahwah, NJ: Erlbaum.

180. Kozlowski, S. W. J., Brown, K. G., Weissbein, D. A., Cannon-Bowers, J. A., & Salas, E., (2000). A multi-level perspective on training effectiveness: Enhancing horizontal and vertical transfer. In: Klein, K. J., & Kozlowski, S. W. J., (eds.), *Multilevel Theory, Research, and Methods in Organizations* (pp. 157–210). San Francisco, CA: Jossey-Bass.

181. Kozlowski, S. W. J., Gully, S. M., Brown, K. G., Salas, E., Smith, E. M., & Nason, E. R., (2001). Effects of training goals and goal orientation traits on multidimensional training outcomes and performance adaptability. *Organ. Behav. Hum. Decis. Process., 85*, 1–31.

182. Kraak, A., (2011). *Cross-Sectoral State Coordination, TVET Policies and Economic Development: Lessons from South Africa Finland, Ireland, and Malaysia.* Background paper commissioned by UNESCO Education Sector.

183. Kraak, A., Jewison, R., Pillay, P., Chidi, M., Bhagwan, N., Sebolelo, N., & Blanche, E., (2014). *Review of the Current Skills Development System and Recommendations Towards the Best Model for Delivering Skills in the Country*. Human Resource Development Council for South Africa.

184. Kraiger, K., (2002). Decision-based evaluation. In: Kraiger, K., (ed.), *Creating, Implementing, and Maintaining Effective Training and Development: State-of-the-Art Lessons for Practice* (pp. 331–375). San Francisco, CA: Jossey-Bass.

185. Kraiger, K., (2003). Perspectives on training and development. In: Borman, W. C., Ilgen, D. R., & Klimoski, R. J., (ed.), *Handbook of Psychology: Industrial and Organizational Psychology* (Vol. 12, pp. 171–192). Hoboken, NJ: Wiley.

186. Kraiger, K., & Aguinis, H., (2001). Training effectiveness: Assessing training needs, motivation, and accomplishments. In: London, M., (ed.), *How People Evaluate Others in Organizations: Person Perception and Interpersonal Judgment in I/O Psychology* (pp. 203–219). Mahwah, NJ: Erlbaum.

187. Kraiger, K., & Jerden, E., (2007). A new look at student control: Meta-analytic results and directions for future research. In: Fiore, S. M., & Salas, E., (eds.), *Where is the Learning in Distance Learning? Towards a Science of Distributed Learning and Training* (pp. 65–90). Washington, DC: APA Books.

188. Kraiger, K., Ford, J. K., & Salas, E., (1993). Application of cognitive, skill-based, and effective theories of learning outcomes to new methods of training evaluation. *J. Appl. Psychol., 78*, 311–328.

189. Kraiger, K., McLinden, D., & Casper, W. J., (2004). Collaborative planning for training impact. *Hum. Resour. Manag. Rev., 43,* 337–351.

190. Latham, G. P., (1988). Human resource training and development. *Annu. Rev. Psychol., 39*, 545–582.

191. Lauterbach, U., (2009). The pedagogy of apprenticeship. In: Maclean, R., & Wilson, D., (eds.), *International Handbook of Education for the Changing World of Work* (pp. 1653–1668). Dordrecht, Springer.

192. Lave, J., & Wenger, E., (1991). *Situated Learning*. Cambridge, Cambridge University Press.

193. Lee, M., (2004). National human resource development in the United Kingdom. *Adv. Dev. Hum. Resour., 6*, 334–345.

194. Leney, T., (2009a). *Qualifications that Count: Strengthening the Recognition of Qualifications in the Mediterranean Region*. Thematic Study. Turin, Italy, ETF. http://www.etf.europa.eu/web.nsf/(RSS)/C125782B0048D6F6C125768200396FB6?OpenDocument&LAN=EN (accessed on 18 June 2022).

195. Leney, T., Ammerman, P., Brandsma, J., Bewhringeer, F., Coles, M., Feenstra, B., Grollmann, P., et al., (2005). *Achieving the Lisbon Goal: The Contribution of VET.* Brussels, European Commission.

196. Lievens, F., Harris, M. M., Van, K. E., & Bisqueret, C., (2003). Predicting cross-cultural training performance: The validity of personality, cognitive ability, and dimensions measured by an assessment center and a behavior description interview. *J. Appl. Psychol., 88*, 476–489.

197. Lim, D. H., & Morris, M. L., (2006). Influence of trainee characteristics, instructional satisfaction, and organizational climate on perceived learning and transfer training. *Hum. Resour. Dev. Q., 17*, 85–115.

198. Linou, N., & Kontogiannis, T., (2004). The effect of training systemic information on the retention of fault-finding skills in manufacturing industries. *Hum. Factors Ergon. Manufacturing, 14*, 197–217.

199. Lippold, K., Nichols, A., & Sorensen, E., (2011). *Strengthening Families Through Stronger Fathers: Final Impact Report for the Pilot Employment Programs*. New York, Urban Institute, 23-30.

200. Littrell, L. N., Salas, E., Hess, K. P., Paley, M., & Riedel, S., (2006). Expatriate preparation: A critical analysis of 25 years of cross-cultural training research. *Hum. Resour. Dev. Rev., 5*, 355–388.

201. Lotz-Sisitka, H., & Raven, G., (2009). South Africa: Applied competence as the guiding framework for environmental and sustainability education. In: Fien, J., Maclean, R., & Park, M. G., (eds.), *Work, Learning and Sustainable Development: Opportunities and Challenges* (pp. 309–318). Dordrecht, Springer and UNEVOC.

202. Mabey, C., & Ramirez, M., (2005). Does management development improve organizational productivity? A six-country analysis of European firms. *Int. J. Hum. Resour. Manag., 16*, 1067–1082.

203. Maclean, R., & Lai, A., (2011). The future of technical and vocational education and training: Global challenges and possibilities. *International Journal of Training Research*, 2–15.

204. Maclean, R., & Pavlova, M., (2011). *Vocationalization of Secondary and Higher Education: Pathways to the World of Work* (pp. 40–85). UNESCO-UNEVOC, Revisiting Global Trends in TVET: Reflections on Theory and Practice. Bonn, Germany, UNESCO-UNEVOC International Centre for Technical and Vocational Education and Training.

205. Maclean, R., Jagannanthan, S., & Sarvi, J., (2011). Main issues, concerns and challenges regarding TVET in Asia and the Pacific. In: Maclean, R., Jagannanthan, S., & Sarvi, J., (eds.), *Skills Development for Inclusive and Sustainable Growth in Developing Asia-Pacific* (pp. 3–27). Manila, Asian Development Bank and Springer.

206. Marks, M. A., Sabella, M. J., Burke, C. S., & Zaccaro, S. J., (2002). The impact of cross-training on team effectiveness. *J. Appl. Psychol., 87*, 3–13.

207. Marock, C., (2010). *Qualifications Frameworks: Implementation and Impact, Background Case Study on Mauritius*. Geneva, ILO Skills and Employability Department.

208. Marope, P. T. M., (2005). *Namibian Human Capital and Knowledge Development for Economic Growth with Equity*. Africa Region Human Development Working Paper Series No. 84. Washington, DC, World Bank.

209. Marope, P. T. M., & Goldberg, I., (2009). *Closing the Skills and Technology Gap in South Africa*. Concept note prepared for a World Bank study.

210. Martin, R., Villeneuve-Smith, F., Marshall, L., & McKenzie, E., (2008). *Employability Skills Explored*. London, Learning and Skills Network (LSN).

211. Martinaitis, Z., (2012). Building on skills forecasts: Comparing methods and applications. *Building on Skills Forecasts: Comparing Methods and Applications; Conference Proceedings, No. 18* (pp. 97–108). Luxembourg, Office for Official Publications of the European Union.

212. Mattson, B., (2005). Using the critical outcome technique to demonstrate financial and organizational performance results. *Adv. Dev. Hum. Resour., 7*, 102–120.

213. McGrath, S., (2011). Skills for development? Rethinking the kind of development, we want TVET to support. In: King, K., (ed.), *Towards a New Global World of Skills Development*? (pp. 8–11). TVET's Turn to Make its Mark. NORRAG News 46.

214. McGrath, S., (2012). Vocational education and training for development. A policy in need of a theory? *International Journal of Educational Development, 32*(5), 623–631.

215. McGrath, S., & Akoojee, S., (2010). Regulating private vocational education and training (VET) in South Africa: The national development imperative. *Africa Education Review, 7*, 16–33.

216. McGrath, S., & Lugg, R., (2012). Knowing and doing vocational education and training reform: Evidence, learning and the policy process. *International Journal of Educational Development, 32*, 696–708.

217. McGrath, S., King, K., Leach, F., & Carr-Hill, R., (1995). *Education and Training for the Informal Sector*. London, Overseas Development Administration.

218. McGuire, D., Cross, C., & O'Donnell, D., (2005). Why humanistic approaches in HRD will not work. *Hum. Resour. Dev. Q., 16*, 131–137.

219. Meek, L. V., (2000). Understanding diversity and differentiation in higher education: An overview. *Higher Education Policy, 13,* 1–6.

220. Mertaugh, M., & Hanushek, E. A., (2005). Education and training. In: Barr, N., (ed.), *Labor Markets and Social Policy in Central and Eastern Europe: The Accession and Beyond* (pp. 207–242). Washington, DC, World Bank.

221. Ministry of Education and Human Resources Development (MEHRD) and KERIS, (2007). *Adapting Education to the Information Age*. Seoul, MEHRD and KERIS.

222. Ministry of Education Finland, (2010). *Noste Program: Final Report* (pp. 24–65). In Finnish.

223. Ministry of Education, Republic of Serbia, (2010). *Vocational Education and Training Reform Program: Teacher Training and Human Resource Development (Belgrade)*. www.vetserbia.edu.rs/humanresourcedevelopment.htm (accessed on 18 June 2022).

224. Mishra, S., & Pulist, S. K., (2013). *eLearning in Commonwealth Asia 2013*. New Delhi, Commonwealth Educational Media Centre for Asia.

225. Misola, N., (2010). *Improving the Participation of Female Students in TVET Programs Formerly Dominated by Males*. Bonn, Germany, UNESCO-UNEVOC International Centre.

226. Morey, J. C., Simon, R., Jay, G. D., Wears, R. L., Salisbury, M., et al., (2002). Error reduction and performance improvement in the emergency department through formal teamwork training: Evaluation results of the MedTeams project. *Health Serv. Res., 37*, 1553–1581.

227. Morgan, R. B., & Casper, W. J., (2000). Examining the factor structure of participant reactions to training: A multidimensional approach. *Hum. Resour. Dev. Q., 11*, 301–317.

228. Morgeson, F. P., & Campion, M. A., (1997). Social and cognitive sources of potential inaccuracy in job analysis. *J. Appl. Psychol., 82*, 627–655.

229. Morris, M. A., & Robie, C., (2001). A meta-analysis of the effects of cross-cultural training on expatriate performance and adjustment. *Int. J. Train. Dev., 5*, 112–125.

230. Murray, F., & Hsi, H., (2007). *Knowledge Workers in Biotechnology: Occupational Structures, Careers and Skill Demands* (pp. 129–137). Paper prepared for the National Academies Workshop on Research Evidence Related to Future Skill Demands.

231. Nielsen, S., (2011). *TVET Teachers and Trainers*. Background paper commissioned by UNESCO Education Sector.

232. Nielsen, S., & Nikolovska, M., (2007). European training foundation. *ETF Yearbook 2007. Quality in Vocational Education and Training*. Luxembourg, Office for Official Publications of the European Communities.

233. Noe, R. A., (2008). *Employee Training and Development* (4th edn.). Boston, MA: Irwin-McGraw.

234. O'Connor, P., Flin, R., & Fletcher, G., (2002). Techniques used to evaluate crew resource management training: A literature review. *Hum. Factors Aerosp. Saf., 2*, 217–233.

235. OECD, (2003). *Beyond Rhetoric: Adult Learning Policies and Practices*. Paris, OECD Publishing.

236. OECD, (2020). *Environmental Policy: Environmental Policy Stringency index*. OECD Environment Statistics (database), https://doi.org/10.1787/2bc0bb80-en.

237. Paradise, A., (2007). *State of the Industry: ASTD's Annual Review of Trends in Workplace Learning and Performance*. Alexandria, VA: ASTD.

238. Phillips, J. J., & Phillips, P., (2007). Measuring return on investment in leadership development. In: Hannum, K. M., Martineau, J. W., & Reinelt, C., (eds.), *The Handbook of Leadership Development Evaluation* (pp. 137–166). Hoboken, NJ: Wiley.

239. Pidd, K., (2004). The impact of workplace support and identity on training transfer: A case study of drug and alcohol safety training in Australia. *Int. J. Train. Dev., 8*, 274–288.

240. Pires, A. L. O., (2010). *Country Report: Portugal*. Torino, CEDEFOP. www.cedefop.europa.eu (accessed on 18 June 2022).

241. Population Reference Bureau, (2013). *The World's Youth 2013 Data Sheet*. Washington, DC, Population Reference Bureau.

242. Psacharopoulos, G., (1981). Returns to education: An updated international comparison. *Comparative Education, 17*, 321–341.

243. Raffe, D., & Spours, K., (2007). Three models of policy learning and policy-making in 1419 education. In: Raffe, D., & Spours, K., (eds.), *Policymaking and Policy Learning in 14–19 Education*. Bedford Way Papers. London, Institute of Education.

244. Rao, T. V., (2004). Human resource development as national policy in India. *Adv. Dev. Hum. Resour., 6*, 288–296.

245. Ravallion, M., (2004). *Pro-Poor Growth: A Primer*. World Bank Policy Research Working Paper No. 3242. Washington, DC, Development Research Group, World Bank.

246. Republic of Tunisia, (2011). *Rapport De La Consultation Nationale Sur L'emploi*. Tunis, Republic of Tunisia.

247. Richman-Hirsch, W. L., (2001). Post-training interventions to enhance transfer: The moderating effects of work environments. *Hum. Resour. Dev. Q., 12*, 105–119.

248. Rivera, R. J., & Paradise, A., (2006). *State of the Industry: ASTD's Annual Review of Trends in Workplace Learning and Performance.* Alexandria, VA: ASTD.

249. Robertson, I. T., (1990). Behavior modeling: Its record and potential in training and development. *Br. J. Manag., 1*, 117–125.

250. Rodrik, D., (2006). *Understanding South Africa's Economic Puzzles.* CID Working Paper No. 130.

251. Rowold, J., (2007). The impact of personality on training-related aspects of motivation: Test of a longitudinal model. *Hum. Resour. Dev. Q., 18*, 9–31.

252. Sackett, P. R., & Mullen, E. J., (1993). Beyond formal experimental design: Towards an expanded view of the training evaluation process. *Pers. Psychol., 46*, 613–627.

253. SACMEQ, (2010). *SACMEQ III Project Results: Pupil Achievement Levels in Reading and Mathematics.* Working Document No. 1. http://www.sacmeq.org/downloads/sacmeqIII/WD01_SACMEQ_III_Results_Pupil_Achievementpdf (accessed on 18 June 2022).

254. SADC and UNESCO, (2011). *Final Report: Status of TVET in the SADC Region* (p. 10). Gaborone, SADC.

255. Salas, E., & Cannon-Bowers, J. A., (2001). The science of training: A decade of progress. *Annu. Rev. Psychol., 52*, 471–499.

256. Salas, E., Burke, C. S., Bowers, C. A., & Wilson, K. A., (2001). Team training in the skies: Does crew resource management (CRM) training work? *Hum. Factors, 42*, 641–674.

257. Salas, E., Wilson, K. A., & Burke, C. S., (2006). Does crew resource management training work? An update, an extension, and some critical needs. *Hum. Factors, 48*, 392–412.

258. Salzano, E., (2005). *Technology-Based Training for Marginalized Girls*. Paris, UNESCO.

259. Satterfield, J. M., & Hughes, E., (2007). Emotion skills training for medical students: A systematic review. *Med. Educ., 41*, 935–941.

260. Schmidt, A., & Ford, J. K., (2003). Learning within a student control training environment: The interactive effects of goal orientation and

metacognitive instruction on learning outcomes. *Pers. Psychol., 56*, 405–430.

261. Sen, A., (1999). *Development as Freedom*. Oxford, Oxford University Press.
262. Seyfried, E., (2011). *Quality Assurance in TVET*. Background paper commissioned by UNESCO Education Sector.
263. Shaikh, N., (2007). *The Present as History: Critical Perspectives on Global Power*. New York, Colombia University Press.
264. Shemyakina, O., (2011). *The Labor Market, Education and Armed Conflict in Tajikistan*. Working Paper 5738. Washington, DC, World Bank.
265. Shiohata, M., (2011). *Skills Acquisition Among Lao Women Textile Weavers in a Changing Economy*. Case study prepared for UNESCO Education Sector.
266. Singh, K., (2012). *Report of the Special Rapporteur on the Right to Education*. Sixty-seventh session of the United Nations General Assembly. Document Number A/67/31. http://www.right-to-education.org/sites/right-to-education.org/files/resource-attachments/UNSR_TVET_2012.pdf (accessed on 18 June 2022).
267. Sirianni, P. M., & Frey, B. A., (2001). Changing a culture: Evaluation of a leadership development program at Mellon financial services. *Int. J. Train. Dev., 5*, 290–301.
268. Sitzmann, T., Bell, B., & Kraiger, K., (2008). A multi-level analysis of the effect of prompting self-regulation in technology-delivered instruction. *Pers. Psychol. Manuscr.* (under review).
269. Sitzmann, T., Brown, K. G., Ely, K., & Kraiger, K., (2008). Motivation to learn in a military training curriculum: A longitudinal investigation. *Mil. Psychol. Manuscr.* (under review).
270. Sitzmann, T., Kraiger, K., Stewart, D., & Wisher, R., (2006). The comparative effectiveness of Web-based and classroom instruction: A meta-analysis. *Pers. Psychol., 59*, 623–664.
271. Smith, E. M., Ford, J. K., & Kozlowski, S. W. J., (1997). Building adaptive expertise: Implications for training design. In: Quinones, M. A., & ˜Ehrenstein, A., (eds.), *Training for a Rapidly Changing Workplace: Applications of Psychological Research* (pp. 89–118). Washington, DC: APA Books.

272. Smith, J., (2008). *Off-Shoring, Outsourcing and the 'Global Labor Arbitrage'* (pp. 180–190). Working Paper. Procida Italy, International Initiative for Promoting Political Economy.
273. Spencer, H., (1859). What knowledge is of most worth? *Westminster Review, 141*, 1–23.
274. Spitzer, D. R., (2005). Learning effectiveness measurement: A new approach for measuring and managing learning to achieve business results. *Adv. Dev. Hum. Resour., 7*, 55–70.
275. Stanley, J., (2013). *Global Validation Workshop on Policies and Practices on TVET Teachers and Instructors in the Arab Region, a Regional Study* (pp. 17–92). Final Report. Paris, UNESCO.
276. Strietska-Ilina, O., Hofmann, C., Haro, D. M., & Shinyoung, J., (2011). *Skills for Green Jobs: A Global View: Synthesis Report Based on 21 Country Studies.* Geneva, International Labor Office, Skills and Employability Department. http://www.ilo.org/wcmsp5/groups/public/---dgreports/---dcomm/---publ/documents/publication/wcms_159585.pdf (accessed on 18 June 2022).
277. Sugrue, B., & Rivera, R. J., (2005). *State of the Industry: ASTD's Annual Review of Trends in Workplace Learning and Performance.* Alexandria, VA: ASTD.
278. Swanson, R. A., (2001). *Assessing the Financial Benefits of Human Resource Development.* Cambridge, MA: Perseus.
279. Sweet, R., (2009). *European Training Foundation, Work-based Learning Programs for Young People in the Mediterranean Region.* Luxembourg, Office for Official Publications of the European Communities.
280. Switzer, K. C., Nagy, M. S., & Mullins, M. E., (2005). The influence of training reputation, managerial support, and self-efficacy on pretraining motivation and perceived training transfer. *Appl. HRM Res., 10*, 21–34.
281. Szalkowski, A., & Jankowicz, D., (2004). The development of human resources during the process of economic and structural transformation in Poland. *Adv. Dev. Hum. Resour., 6*, 346–354.
282. Tannenbaum, S. I., & Woods, S. B., (1992). Determining a strategy for evaluating training: Operating within organizational constraints. *Hum. Resour. Plann., 15*(2), 63–81.
283. Tannenbaum, S. I., & Yukl, G., (1992). Training and development in work organizations. *Annu. Rev. Psychol., 43*, 399–441.

284. Taylor, P. J., Russ-Eft, D. F., & Chan, D. W. L., (2005). A meta-analytic review of behavior modeling training. *J. Appl. Psychol., 90*, 692–709.

285. TESDA, (2015). *SICAT – Trainers Methodology Level 1 Online Training Program*. www.sicattesda.ph (accessed on 18 June 2022).

286. Tharenou, P., Saks, A. M., & Moore, C., (2007). A review and critique of research on training and organizational-level outcomes. *Hum. Resour. Manag. Rev., 17*, 251–273.

287. Towler, A. J., (2003). Effects of charismatic influence training on attitudes, behavior, and performance. *Pers. Psychol., 56,* 363–381.

288. Tracey, J. B., Hinkin, T. R., Tannenbaum, S., & Mathieu, J. E., (2001). The influence of individual characteristics and the work environment on varying levels of training outcomes. *Hum. Resour. Dev. Q., 12*, 5–23.

289. Trewin, et al., (2019). Headline indicators for global climate monitoring. Bulletin of the American Meteorological Society. Accepted for publication. https://doi.org/10.1175/BAMS-D-19–0196.1.

290. Trow, M., (1995). *Diversity in Higher Education in the United States of America*. Paper presented at the CVCP Seminar on Diversity in Higher Education. London.

291. Tuck, R., (2007). *An Introductory Guide to National Qualifications Frameworks: Conceptual and Practical Issues for Policy Makers*. Geneva, ILO Skills and Employability Department.

292. Twitchell, K., Holton, E. III., & Trott, J. W. Jr., (2000). Technical training evaluation practices in the United States. *Perform. Improv. Q., 13*(1), 84–109.

293. Ubeda, G. M., (2005). Training and business performance: The Spanish case. *Int. J. Hum. Resour. Manag., 16*, 1691–1710.

294. UNCTAD, (2011). *Foreign Direct Investments in LDCs: Lessons Learned from the Decade 2001–2010 and the Way Forward*. Geneva, United Nations Conference on Trade and Development.

295. UNDP, (2009). *Capacity Development: A UNDP Primer* (pp. 56–71). New York, UNDP.

296. UNEP, (2008). *Green Jobs: Towards Decent Work in a Sustainable, Low Carbon World* (pp. 39–53). New York, UNEP.

297. UNEP, (2011). *Towards a Green Economy: Pathways to Sustainable Development and Poverty Eradication*. http://www.unep.org/greeneconomy/ (accessed on 18 June 2022).

298. UNESCO-Bangkok, (2011). *TVET in Asia and the Pacific Region: Overview of the Regional Background Paper on TVET: Asia and the Pacific Region.* Unpublished background paper.

299. UNESCO-UNEVOC, (2004). *Bonn Declaration on Learning for Work, Citizenship and Sustainability*. UNESCO international meeting of technical and vocational education and training experts. Bonn, Germany. http://www.unevoc.unesco.org/fileadmin/user_upload/pubs/SD_BonnDeclaration_e.pdf (accessed on 18 June 2022).

300. UNESCO-UNEVOC, (2012). *UNEVOC in Action, Biennial Report 2010–2011*. Bonn, Germany, UNESCO-UNEVOC International Center for Technical and Vocational Education and Training.

301. UNESCO, (1989). *Convention on Technical and Vocational Education*. Paris, UNESCO. http://portal.unesco.org/en/ev.php-URL_ID=13059&URL_DO=DO_TOPIC&URL_SECTION=201.html (accessed on 18 June 2022).

302. UNIDO, (2009). *Industrial Development Report 2009: Breaking in and Moving Up: New Industrial Challenges for the Bottom Billion and the Middle-Income Countries*. Vienna, UNIDO.

303. United Nations (2015). "*THE 17 GOALS | Department of Economic and Social Affairs.*" https://sdgs.un.org/goals (accessed on 18 June 2022).

304. United Nations, (1992). *Report of the United Nations Conference on Environment and Development* (A/CONF.151/26 Vol.1), Rio de Janeiro, UN General Assembly.

305. UNSTT, (2012). *Realizing the Future, we Want for All* (pp. 45–60). Report to the Secretary-General. New York, United Nations System Task Team.

306. Valente, C., (2011). *What Did the Maoists Ever do for us? Education and Marriage of Women Exposed to Civil Conflict in Nepal*. Working Paper 5741.Washington, DC, World Bank.

307. Van, D. K. M., Gielen, E., & Nauta, C., (2001). Supervisory support as a major condition to enhance transfer. *Int. J. Train. Dev., 5*, 52–63.

308. Van, L. M. J., & Van, P. B. M., (2002). The costs and benefits of lifelong learning: The case of the Netherlands. *Hum. Resour. Dev. Q., 13*, 151–168.

309. Vancouver, J. B., & Day, D. V., (2005). Industrial and organization research on self-regulation: From constructs to applications. *Appl. Psychol., 54*, 155–185.

310. Varghese, N., & Puttmann, V., (2011). *Trends in Diversification of Post-Secondary Education*. Paris, UNESCO-IIEP.

311. Veal, K., (2008). *Once Were Warriors: Reintegrating Ex-Combatants*. Supplement to UNESCO- UNEVOC Bulletin, No. 14.

312. Walther, R., (2007). *Vocational Training in the Informal Sector*. Paris, Agence Française de Développement.

313. Wang, G. G., & Wilcox, (2006). Training evaluation: Knowing more than is practiced. *Adv. Dev. Hum. Resour., 8*, 528–539.

314. Wang, G. G., Dou, Z., & Li, N., (2002). A systems approach to measuring return on investment for HRD interventions. *Hum. Resour. Dev. Q., 13*, 203–224.

315. Wang, G., (2002). Control groups for human performance technology (HPT) evaluation and measurement. *Perform. Improv. Q., 15*(2), 34–48.

316. Watts, A. G., & Fretwell, D., (2004). *Public Policies for Career Development*. Washington, DC, World Bank.

317. Welsh, E. T., Wanberg, C. R., Brown, K. G., & Simmering, M. J., (2003). E-learning: Emerging uses, empirical results, and future directions. *Int. J. Train. Dev., 7*, 245–258.

318. Wesson, M. J., & Gogus, C. I., (2005). Shaking hands with a computer: An examination of two methods of organizational newcomer orientation. *J. Appl. Psychol., 90*, 1018–1026.

319. Wexley, K. N., (1984). Personnel training. *Annu. Rev. Psychol., 35*, 519–551.

320. Wilson, R., & Zuckersteinova, A., (2011). *Anticipating Changing Skills Needs: A Master Class*. Technical Reference Document for New Skills. University of Warwick IER/CEDEFOP.

321. WMO Statement on the State of the Global Climate in 2019. (2020). World Meteorological Organization. https://library.wmo.int/doc_num.php?explnum_id=10211 (accessed on 18 June 2022).

322. Woods, P., (2004). *Environmental Education for a Sustainable Future: Formal Schooling*. Curriculum leadership. http://www.curriculum.edu.au/leader/environmental_education_for_a_sustainable_future:_,4695.html (accessed on 18 June 2022).

323. World Bank and DfID, (2003). *Knowledge and Skills in the New Economy: Knowledge and Skills in the MENA Region, Tunisia*. Washington, DC, World Bank.

324. World Bank, (2005a). *World Development Report 2006: Equity and Development*. Washington, DC, World Bank.

325. World Commission on Environment and Development, (1987). *Our Common Future*. Oxford, Oxford University Press.

326. World Conference on Education for All, (2000). *The Dakar Framework for Action: Meeting our Collective Commitments to Education for All*. Paris, UNESCO.

327. World Economic Forum, (2008). *The Global Competitiveness Report 2008–2009* (pp. 34, 35). Geneva, World Economic Forum.

328. World Resources Institute, (2011). *A Compilation of Green Economy Policies, Programs, and Initiatives from Around the World* (pp. 12–19). The Green Economy in Practice: Interactive Workshop 1.

329. Yang, J., & Valdes-Cotera, R., (2011). *Conceptual Evolution and Policy Development in Lifelong Learning*. Bonn, Germany, UNESCO Institute for Lifelong Learning.

330. Yarnit, M., (2004). *Regeneration and All That: Learning and Skills and Sustainability*. London, Learning and Skills Development Agency.

331. Zhao, Y., Lei, J., Lai, B. Y. C., & Tan, H. S., (2005). What makes the difference? A practical analysis of research on the effectiveness of distance education. *Teach. Coll. Rec., 107*, 1836–1884.

INDEX

Printed in the United States
by Baker & Taylor Publisher Services